Eberhard Freitag

Riemann Surfaces

Sheaf Theory, Riemann Surfaces, Automorphic Forms

Self-Publishing, 2014

Eberhard Freitag
Universität Heidelberg
Mathematisches Institut
Im Neuenheimer Feld 288
69120 Heidelberg
`freitag@mathi.uni-heidelberg.de`

This work is subject to copyright. All rights are reserved.
© Self-Publishing, Eberhard Freitag

Contents

Chapter I. Topological spaces   8

1. The notion of a topological space   8
2. Continuous maps   9
3. Special classes of topological spaces   11
4. Sequences   12
5. Connectedness   13
6. Paracompactness   13
7. Frèchet spaces   15

Chapter II. Some algebra   19

1. Abelian groups   19
2. Some homological algebra   20
3. The tensor product   23

Chapter III. Sheaves   26

1. Presheaves   26
2. Germs and Stalks   27
3. Sheaves   29
4. The generated sheaf   31

Chapter IV. Cohomology of sheaves   34

1. The canonical flabby resolution   34
2. Sheaves of rings and modules   39
3. Čech Cohomology   42
4. Some vanishing results   46

Chapter V. Basic facts about Riemann surfaces   50

1. Geometric spaces   50
2. The notion of a Riemann surface   51
3. The notion of a differentiable manifold   53
4. Meromorphic functions   54
5. Ramification points   56
6. Examples of Riemann surfaces   60
7. Differential forms   64

Chapter VI. The Riemann–Roch theorem   72

1. Generalities about vector bundles   72
2. The finiteness theorem   76
3. The Picard group   79
4. Riemann–Roch   81
5. A residue map   86
6. Serre duality   89
7. Some comments on the Riemann–Roch theorem   93

Chapter VII. The Jacobi inversion problem   97

1. Harmonic differentials   97
2. Hodge theory of compact Riemann surfaces   99
3. Integration of closed forms and homotopy   102
4. Periods   103
5. Abel's theorem   108
6. The Jacobi inversion problem   112
7. The fibres of the Jacobi map   116
8. The Jacobian inversion problems and abelian functions   117

Chapter VIII. Dimension formulae for automorphic forms   121

1. Fuchsian groups and Riemann surfaces   122
2. Vector valued automorphic forms   127
3. Direct and inverse images   135
4. The computation of the degree, first method   139
5. The dimension formula   144
6. The full modular group   145
7. The metaplectic group   147

Appendix. Generalizations of the dimension formula	149
8. Chern forms	149
9. The computation of the degree, second method	152
Literature	158
Index	159

Introduction

In many textbooks about Riemann surfaces, the theorem of Riemann–Roch for divisors or line bundles has been treated. But usually they do not contain the theorem for vector bundles. Here the reader usually is referred to the Hirzebruch–Riemann–Roch theorem. The Riemann–Roch theorem for vector bundles has very elementary applications. For example, one can use it to compute dimensions of vector spaces of vector valued elliptic modular forms as they are used for example in the theory of Borcherds products.

One purpose of this book is to give a quick proof of the Riemann–Roch theorem for vector bundles on a compact Riemann surface, and, as an application, we will deduce these dimension formulae in a rather general context.

The approach to Riemann surfaces is sheaf theoretic, quite different than in our textbook [Fr1]. There are similarities with the important book of Forster [Fo]. Forster also uses the sheaf theoretic approach. To keep this as elementary as possible, he restricts to introduce the first Čech cohomology group of a sheaf. Instead of this we introduce a full cohomology theory of sheaves in a form which allows to use this approach also in other mathematical topics as for example in algebraic geometry. Actually, we use Godement's canonical flabby resolution to introduce the cohomology groups of a sheaf.

In the one variable case, the Riemann–Roch theorem is just a consequence of the finiteness theorem which states that the cohomology groups $H^i(X, \mathcal{M})$ of a vector bundle $\mathcal{M}$ on a compact Riemann surface are finite dimensional and vanish for $i > 1$.

The reduction of the Riemann–Roch theorem to the finiteness theorem gets particularly transparent if one formulates and proves it not only for vector bundles but for coherent sheaves. It is not necessary to introduce here the general machinery of coherent sheaves. Coherent sheaves can be introduced as extensions of vector bundles by skyscraper sheaves.

This version of the Riemann–Roch theorem does not contain the duality theorem. As usual in the modern approaches, its proof is separated from the proof of the Riemann–Roch theorem and needs some extra effort as a cohomological version of the residue.

After the proof of the Riemann–Roch theorem we treat the Jacobi inversion problem in an extra Chapter VII. We start this chapter with the Hodge theory of Riemann surfaces which shows that the first cohomology group with coefficients in $\mathbb{C}$ can be described by means of harmonic differentials. Readers who are mainly interested in the mentioned dimension formulae can skip this chapter and switch to the last chapter where these formulae are derived.

In the last Chapt. VIII we derive the mentioned dimension formulae. First we give a very short introduction to discrete subgroups of $\mathrm{SL}(2,\mathbb{R})$ and the associated Riemann surfaces including their extension by cusps. We are interested in the case where these surfaces are compact. Then we introduce vector valued automorphic forms which belong to real weight and matrix valued multiplier systems. We use the Riemann–Roch theorem to derive basic dimension formulae for spaces of automorphic forms. We describe two different methods for the computation of the degree of the occurring vector bundles. The first one works for rational weight and multiplier systems which get trivial on a subgroup of finite index. This method is very simple and needs no further ingredients. For practical purposes, this special case is sufficient. But we don't want to forget about the general case completely and prove the dimension formulae also in the general case in a compendious appendix. Here we have to introduce the notion of the Chern form of a line bundle and the formula of Gauss-Bonnet comes into the game.

Chapter I. Topological spaces

1. The notion of a topological space

A topology $\mathcal{T}$ on a set X is a system of subsets – called open subsets – with the following properties:

1) $\emptyset$, X are open.
2) The intersection of finitely many open subsets is open.
3) The union of arbitrarily many open sets is open.

A *topological space* is a pair $(X, \mathcal{T})$, consisting of a set X and a topology $\mathcal{T}$ on X. Usually we will write X instead of $(X, \mathcal{T})$ when it is clear which topology is considered. We give some constructions for topological spaces:

Metric spaces

Let (X, d) be a metric space. We denote by $U_r(a)$ the ball around a of radius r. A subset U of X is called open if for every $a \in U$ there exists $\varepsilon > 0$ with the property $U_\varepsilon(a) \subset U$.

The induced topology

Let Y be a subset of a topological space $X = (X, \mathcal{T})$. Then Y can be equipped with the *induced topology* $\mathcal{T}|Y$.

A subset $V \subset Y$ belongs to $\mathcal{T}|Y$ if and only if there exists a subset $U \subset X$, $U \in \mathcal{T}$, such that
$$V = U \cap Y.$$
(When Y is an open subset of X then this means $V \in \mathcal{T}$.) Since a subset $V \subset Y$ is also a subset of X, one has to make clear whether "open" means to be in $\mathcal{T}$ or in $\mathcal{T}|Y$. If we say "open in X" we mean that V is contained in $\mathcal{T}$. Similarly "open in Y" means that it is contained in $\mathcal{T}|Y$.

The quotient topology

Let X be a topological space and let $f : X \to Y$ be a surjective map onto a set Y. Then Y can be equipped with the *quotient topology*. A subset $V \subset Y$ is called open if and only if the inverse image $U := f^{-1}(V)$ is open in X. There is an important special case: Let „$\sim$" be an equivalence relation on X, let $Y = X/\sim$ be the set of equivalence classes, and let $f : X \to Y$ be the canonical projection. Then Y is called the quotient space of X. Examples are a torus $X = \mathbb{C}/L$ for a lattice L or the modular space $\mathbb{H}/\operatorname{SL}(2, \mathbb{Z})$.

The product topology

Let $X_1, \ldots, X_n$ be a finite system of topological spaces. Then a natural topology on the cartesian product

$$X = X_1 \times \cdots \times X_n$$

can be defined.

A subset $U \subset X$ is called open if and only if for every point $a \in U$ there exist open subsets $U_1 \subset X_1, \ldots, U_n \subset X_n$, such that

$$a \in U_1 \times \cdots \times U_n \subset U.$$

2. Continuous maps

A subset of a topological space $A \subset X$ is called *closed* if the complement $X - A$ is open.

A subset $M \subset X$ is called a *neighborhood* of a point $a \in X$ if there exists an open subset $U \subset X$ with $a \in U \subset M$.

A point $a \in X$ is called a *boundary point* of a subset $M \subset X$ if every neighborhood of a contains points of M and of its complement $X - M$.

Notation.
$$\partial M := \text{set of boundary points},$$
$$\bar{M} := M \cup \partial M.$$

One shows easily that $\bar{M}$ is the smallest closed subset of X which contains M. In particular,
$$M \text{ closed} \iff M = \bar{M}.$$
We call $\bar{M}$ the *closure* of M. Similarly, there exists a biggest open subset M° of a set M. We call it the *interior* of M.

A map $f : X \to Y$ between topological spaces is called *continuous* at a point $a \in X$ if the inverse image $f^{-1}(V(b))$ of an arbitrary neighborhood of $b := f(a)$ is a neighborhood of a.

The following conditions are equivalent.

1) The map f is continuous (i.e. continuous at every point).
2) The inverse image of an arbitrary open subset of Y is open in X.
3) The inverse image of an arbitrary closed subset of Y is closed in X.

The composition of two continuous maps is continuous. (This is true already in the pointwise sense.)

Some universal properties

Let Y be a subset of a topological space equipped with the induced topology. A map $f : Z \to Y$ of a third topological space into Y is continuous if and only if the composition with the natural inclusion $Y \hookrightarrow X$ is a continuous map $Z \to X$.

Let $f : X \to Y$ be a surjective map of topological spaces, where Y carries the quotient topology. A map $Y \to Z$ into a third topological space Z is continuous if and only if the composition with f is a continuous map $X \to Z$.

Let $X_1, \ldots, X_n$ be topological spaces and let

$$f : Y \longrightarrow X_1 \times \cdots \times X_n$$

be a map of another topological space Y into the cartesian product (equipped with the product topology). The map f is continuous if and only if each component

$$f_j : Y \longrightarrow X_j \qquad (f = (f_1, \ldots, f_n))$$

is continuous.

Topological maps

A map $f : X \to Y$ between topological spaces is called *topological* if it is bijective and if f and f^{-1} both are continuous. Two topological spaces X, Y are called *topologically equivalent* or *homeomorphic* if there exists a topological map between them.

For example, the 2-sphere

$$S^2 = \left\{ x \in \mathbb{R}^3;\ x_1^2 + x_2^2 + x_3^2 = 1 \right\}$$

and the *Riemann sphere* are homeomorphic. Here the Riemann sphere $\bar{\mathbb{C}} = \mathbb{C} \cup \{\infty\}$ is equipped with the following topology. A subset $U \subset \bar{\mathbb{C}}$ is open if the following two conditions are satisfied:

a) $U \cap \mathbb{C}$ is open.
b) If $\infty \in U$ then there exists $C > 0$ such that $\{z \in \mathbb{C};\ |z| > C\} \subset U$.

A topological map $S^2 \to \bar{\mathbb{C}}$ can be constructed by means of the stereographic projection.

If $L \subset \mathbb{C}$ is a lattice, then $\mathbb{C}/L$ is homeomorphic to the cartesian product of two circles:

$$\boxed{\mathbb{C}/L = \text{Torus} \simeq S^1 \times S^1.}$$

3. Special classes of topological spaces

A topological space X is called *Hausdorff* if for two different points $a, b \in X$ there exist disjoint neighborhoods $U(a)$ and $U(b)$.

A topological space X is called *compact* if it is Hausdorff and if every open covering admits a finite sub-covering. A subset Y of a topological space X is called compact if it is – equipped with the induced topology – a compact topological space.

Properties of compact subsets

a) Compact subsets are closed.
b) A closed subset of a compact space is compact.
c) Let $f : X \to Y$ be a continuous map of Hausdorff spaces, then the image of a compact subset is compact.
d) Let X be a compact, let Y a Hausdorff space, and let $f : X \to Y$ bijective and continuous. Then f is topological.
e) The product $X_1 \times \cdots \times X_n$ of compact spaces is compact.

A topological space is called *locally compact* if it is Hausdorff and if every point admits a compact neighborhood. The product $X_1 \times \cdots \times X_n$ of locally compact spaces is locally compact.

Proper maps

A continuous map $f : X \to Y$ of Hausdorff spaces is called *proper* if the inverse image of an arbitrary compact subset is compact. If Y consists only of one point then this means that X is a compact space. Hence "proper" should be considered as a relative version of compact. A simple example of a proper map is

$$\mathbb{C} \longrightarrow \mathbb{C}, \quad z \longmapsto z^n,$$

for natural numbers n.

3.1 Lemma. *Let $f : X \to Y$ be a proper map. Let $B \subset Y$ be a subset and $A = f^{-1}(B)$ the full inverse image of f. Then the restriction $A \to B$ is also proper.*

Proof. Let $K \subset B$ be compact. Then the inverse image of K considered in X and in A is the same, since A is the full inverse image. □

4. Sequences

A sequence (a_n) in a topological space X converges to $a \in X$ if for every neighborhood $U(a)$ there exists $N \in \mathbb{N}$ such that $a_n \in U$ for $n \geq N$. We simply write $a_n \to a$ for this. If X is Hausdorff, the limit a is unique. If $f : X \to Y$ is continuous at a, then $a_n \to a$ implies $f(a_n) \to f(a)$.

It is easy to show that in a compact space every sequence admits a convergent sub-sequence. The converse is only true under additional conditions.

One says that a topological space has a *countable* basis of the topology if there exists a countable family (U_i) of open subsets such that every open subset can be written as union of members of this system. If this is the case, also every subset equipped with the induced topology has countable basis of the topology. The space $\mathbb{R}^n$ with the usual topology has countable basis. One can take open balls where the radius and the coordinates of the center are rational numbers.

4.1 Proposition. *Assume that X is a Hausdorff space such that every sequence admits a convergent sub-sequence. Then X is compact if one of the following two conditions is satisfied.*

a) *The topology comes from a metric.*
b) *There exists a countable basis of the topology.*

Let X be a locally compact space with countable basis $(U_i)_{i \in I}$ of the topology. If we take the sub-system $(U_i)_{i \in J}$, $J \subset I$ of all U_i with compact closure, then this is still a basis of the topology in the sense that every open subset is a union of members of this system. Indeed, the open subsets U with compact closures can be covered by members of the restricted system and, since X is locally compact, every open subset can be covered by open subsets with compact closure. Hence we can write $X = U_1 \cup U_2 \cup \ldots$ as union of a sequence of open subsets with compact closure. Replacing inductively U_n by the union of the U_i with $i < n$, we can get $U_1 \subset U_2 \subset \ldots$. For given n, the closure of U_n can be covered by finitely many of the U_i. Hence, taking a suitable subsequence we can obtain that the closure of U_n is contained in U_{n+1}. Hence we have obtained the following result.

4.2 Lemma. *Let X be a locally compact space with countable basis of the topology. Then X can be written as union of countable many compact subsets,*

$$X = K_1 \cup K_2 \cup \ldots.$$

We can assume that K_n is contained in the interior of K_{n+1}.

5. Connectedness

A topological space is called *arcwise connected* if every two points are contained in (the image of) an arc. An arc is a continuous map of a real interval into X.

A topological space is called *connected* if one of the following two equivalent conditions is satisfied.

1) Every locally constant map $f : X \to M$ into an arbitrary set M is constant. It is sufficient to verify this for one set M which contains at least two elements.
2) If $X = U \cup V$ is written as union of two disjoint open subsets U, V, then one them is empty.

The mean value theorem of calculus shows that every real interval is connected. As a consequence, every arcwise connected space is connected. The converse is only true under additional assumptions (s. below).

Arc components

Two points of a topological space are called equivalent if there exists an arc which contains both. The equivalence classes of this equivalence relation are called *arc components*.

A *topological manifold* is a Hausdorff space such that every point admits an open neighborhood which is homeomorphic to some open subset of $\mathbb{R}^n$. If n can be taken to be two, then X is called a *(topological) surface*.

5.1 Remark. *Let X be a topological manifold. Then the arc components are open. The manifold is connected if and only if it is arcwise connected.*

Since an open subset of a manifold is a manifold, the arc components of a manifold are also manifolds. Of course they are connected. We call the arc components of a manifold also the *connected components*.

6. Paracompactness

We recall the definition and basic properties of *paracompact* spaces. The proofs can be found in standard text books of topology, as for example [Qu].

A covering $\mathfrak{U} = (U_i)_{i \in I}$ of a topological space is called *locally finite* if for every point $a \in X$ there exists a neighborhood W, such that the set of indices $i \in I$ with $U_i \cap W \neq \emptyset$ is finite.

A covering $\mathfrak{V} = (V_j)_{j \in J}$ is called a *refinement* of the covering $\mathfrak{U}$ if for every index $j \in J$ there exists an index $i \in I$ with $V_j \subset U_i$. If one chooses for each j such an i, one obtains a so-called *refinement map* $J \to I$ which needs not to be unique.

6.1 Definition. *A Hausdorff space is called* **paracompact** *if every open covering admits a locally finite open refinement.*

Every compact space is paracompact. We need the following result.

6.2 Lemma. *Every locally compact space with countable basis of the topology is paracompact.*

We sketch the simple proof. We use Lemma 4.2 and write $X = K_1 \cup K_2 \cup \ldots$ as union of a sequence of compact sets such that K_n is contained in the interior K_{n+1}° of K_{n+1}. We set $K_0 = \emptyset$. Now we consider an open covering (U_i) of X. For each natural number n we cover $K_n - K_{n-1}^\circ$ by the open sets $U_i \cap (K_{n+1}^\circ - K_{n-1})$. There exists a finite subsystem which covers K_n. If we collect these sets for all n we get a refinement of (U_i) which is locally finite. □

Just for sake of completeness we mention that each metric space is paracompact and that every locally compact space with countable basis of the topology can be metricised.

Next we formulate the basic result about paracompactness. Let $\mathfrak{U} = (U_i)$ be a locally finite open covering. A *partition of unity* with respect to $\mathfrak{U}$ is a family φ_i of continuous real valued functions on X with the following propertie.s

a) The support of φ_i is contained in U_i.
b) $0 \le \varphi_i \le 1$.
c) $\sum_{i \in I} \varphi_i(x) = 1$ for all $x \in X$.

(This sum is finite for each x. Recall that the support is the closure of the set of all points where the function is different form zero.)

6.3 Proposition. *Let X be a paracompact space. For every locally finite open covering there exists a partition of unity.*

A proof can be found in many textbooks of topology, for example in [Qu]. We mention that the proof gets very simple if X is a Hausdorff topological manifold with a countable basis of topology. This is enough for our purposes.

We mention two related results.

6.4 Proposition. *Let X be a paracompact space and $\mathfrak{U} = (U_i)$ a locally finite open covering. There exist open subsets $V_i \subset U_i$ whose closure $\bar{V}_i$ (taken in X) is contained in U_i and such that $\mathfrak{V} = (V_i)$ is still a covering.*

Let $V \subset U$ are two open subsets of a paracompact space such the closure $\bar{V}$ of V (taken in X) is contained U. Then we can consider the open covering $X = U \cup (X - \bar{V})$. By means of a partition of unity with respect to this covering we obtain the following result.

§7. Frèchet spaces

6.5 Proposition. *Let X be a locally compact paracompact space, U an open subset and $V \subset U$ an open subset whose closure (taken in X) is contained in U. Then there exists a continuous function on X which is one on V and whose support is contained in U.*

7. Frèchet spaces

A topological vector space is a complex vector space V equipped with a topology such that the maps

$$V \times V \longrightarrow V, \ (a, b) \longmapsto a + b, \qquad \mathbb{C} \times V \longrightarrow X, \ (C, a) \longmapsto Ca,$$

are continuous. Here $V \times V$ and $\mathbb{C} \times V$ carry of course the product topology. For example $\mathbb{C}^n$ equipped with the usual topology is a topological vector space. More generally, every finite dimensional vector space V can be equipped in a unique way with a structure as topological space such that each isomorphism $V \to \mathbb{C}^n$ is a topological map.

In a topological vector space one can talk about *Cauchy sequences*. A sequence (a_n) is called a Cauchy sequence if for every neighbourhood U of 0 there exists N such that $a_m - a_n \in U$ for all $m, n \geq N$. Every convergent sequence is a Cauchy sequence. A topological vector space is called *complete* if every Cauchy sequence converges.

A semi-norm p on a complex vector space V is a map $p : V \to \mathbb{R}$ with the properties

a) $p(a) \geq 0$ for all $a \in V$,
b) $p(ta) = |t|p(a)$ for all $t \in \mathbb{C}, \ a \in V$,
c) $p(a + b) \leq p(a) + p(b)$.

The *ball* of radius $r > 0$ is defined as

$$U_r(a, p) := \{\, x \in V;\ p(a - x) < r \,\}.$$

Let $\mathcal{M}$ be a set of semi-norms. A subset $B \subset V$ is called a *semi-ball* around a with respect to $\mathcal{M}$ if there exists a finite subset $\mathcal{N} \subset \mathcal{M}$ such that B is the intersection of balls with respect to the elements of $\mathcal{N}$

$$B = \bigcap_{p \in \mathcal{N}} U_{r(p)}(a, p).$$

A subset U of V is called open (with respect to $\mathcal{M}$) if for every $a \in U$ there exists a semi-ball B around a with $B \subset U$.

It is clear that this defines a topology on V and that V gets a topological vector space. The semi-balls are open. All maps $p : V \to \mathbb{C}$ are continuous and the topology is actually the weakest topology with this property.

A sequence (a_n) in V converges to $a \in V$ if and only if $p(a_n - a) \to 0$ for all $p \in \mathcal{M}$.

A sequence (a_n) in V is a *Cauchy sequence* if for every $\varepsilon > 0$ and if for every $p \in \mathcal{M}$ there exists an $N = N(p, \varepsilon)$ such that

$$p(a_n - a_m) < \varepsilon \quad \text{for} \quad n, m \geq N.$$

The set $\mathcal{M}$ is called *definite* if

$$p(a) = 0 \quad \text{for all} \quad p \in \mathcal{M} \quad \Longrightarrow \quad a = 0.$$

It is easy to prove that V is a Hausdorff space if and only if $\mathcal{M}$ is definite.

7.1 Definition. *A Frèchet space V is a Hausdorff topological vector space with the following property. It is complete and the topology can be defined by means of a **countable** set of semi-norms on V.*

We formulate a simple remark.

7.2 Remark. *A closed subspace of a Frèchet space is a Frèchet space. The direct product of finitely many Frèchet spaces is a Frèchet space.*

The proof is very easy. To construct the semi-norms for a product, one uses the construction $\max(p_1(a_1), \ldots, p_n(a_n))$. The details can be left to the reader. □

We recall that a semi-norm p is called a *norm* if it is definite in the sense that $p(a) = 0$ implies $a = 0$. A *normed space* is a vector space with a distinguished norm. It is called a *Banach space* if it is complete with respect to this norm. So Frèchet spaces can be considered as generalizations of Banach spaces.

7.3 Lemma. *Frèchet spaces are metrizable.*

Proof. We choose a countable set of defining semi-norms $\mathcal{M} = \{p_1, p_2, \ldots\}$. Then one defines

$$d(a, b) = \sum_{n=1}^{\infty} 2^{-n} \frac{p_n(a-b)}{1 + p_n(a) + p_n(b)}.$$

It is easy to show that this is a metric which defines the original topology. □

As a consequence, a subset M of a Frèchet space is compact if and only if each sequence in M has a convergent sub-sequence with limit in M.

§7. Frèchet spaces

Basic examples of Frèchet spaces

Let $U \subset \mathbb{C}$ be an open subset and $\mathcal{O}(U)$ the set of all analytic functions on U. This is a complex vector space. For an arbitrary compact subset $K \subset U$ we define
$$p(f) = p_K(f) := \max_{z \in K} |f(z)|.$$
This is s semi norm. (Actually, it is a norm if U is connected and if K contains a non-empty open subset of U.) A sequence (f_n) converges with respect to p_K if and only if f_n converges uniformly on K.

7.4 Remark. *Let $U \subset \mathbb{C}$ be an open subset. The vector space $\mathcal{O}(U)$ equipped with the set of all norms of the form p_K, $K \subset U$ compact, is a Frèchet space.*

Proof. Recall that U can be exhausted by a sequence of compact subsets K_n such that K_n is contained in the interior of K_{n+1}. It is enough to take for K the members of this sequence. Hence the topology can be defined by means of a countable set of semi-norms. The convergence of Cauchy sequences follows from the theorem of Weierstrass, which states that analyticity is stable under uniform convergence on compact subsets. □

The same argument shows that the space of continuous functions $\mathcal{C}(U)$ is a Frèchet space. The basic result about the Frèchet space $\mathcal{O}(U)$ is the following.

7.5 Theorem of Montel. *Let U be an open subset of $\mathbb{C}$ and $C > 0$ a positive constant. The set*
$$\mathcal{O}(U, C) := \{ f \in \mathcal{O}(U);\ |f(z)| \leq C \text{ for } z \in U \}$$
is compact in $\mathcal{O}(U)$.

For the proof one has to use the fact that a metric space is compact if every sequence admits a convergent subsequence. Hence the statement follows from the usual theorem of Montel which states that every sequence in $\mathcal{O}(U, C)$ admits a locally convergent sub-sequence. A proof can be found in [FB], Kap. IV, Theorem 4.9. We notice that the analogue for differentiable functions is false. The proof of Theorem 7.5 uses heavily the Cauchy integral.

Compact operators

A well-known fact is that in a Banach space of infinite dimension the closed ball $||a|| \leq 1$ is not compact. This result is also true for Frèchet spaces in the following form:

Assume that the Frèchet space admits a non-empty open subset with compact closure. Then it is of finite dimension.

We need a generalization of this result. A continuous linear map $f : E \to F$ between Frèchet spaces is called a *compact operator* if there exists a non-empty open subset of E such that the closure of its image is compact.

A linear map $f : V \to W$ is called *nearly surjective* if $W/f(V)$ has finite dimension. This is automatically the case when W is finite dimensional.

7.6 Theorem of Schwartz. *Let $f : E \to F$ be a surjective continuous linear map between Frèchet spaces and let $g : E \to F$ be a compact operator. Then $f + g$ is nearly surjective.*

A short proof can be found in [GR], Appendix B, Theorem 12.

If one applies Schwartz's theorem in the case $E = F$, $f = -\operatorname{id}$, and $g = \operatorname{id}$, one obtains the following result.

7.7 Corollary. *When the identity operator $\operatorname{id} : E \to E$ of a Frèchet space is compact, then E is finite dimensional.*

From Montel's theorem we obtain the following important example of a compact operator.

7.8 Proposition. *Let $U \subset \mathbb{C}$ be an open subset and $V \subset U$ an open subset whose closure (taken in $\mathbb{C}$) is compact and contained in U. The natural restriction map $\mathcal{O}(U) \to \mathcal{O}(V)$ is compact.*

Proof. Consider $K = \bar{V}$ and the semi-norm p_K on $\mathcal{O}(U)$. The set of all functions $f \in \mathcal{O}(U)$ with $p_K(f) < 1$ is open. Its image in $\mathcal{O}(V)$ is contained in a compact set by Montel's theorem.

Chapter II. Some algebra

1. Abelian groups

We assume that the reader is familiar with the notion of an abelian group and homomorphism between abelian groups. If A is a subgroup of an abelian group B, then the factor group B/A is well defined. All what one needs usually is that there is a natural surjective homomorphism $f : B \to B/A$ with kernel A. Let $f : B \to X$ be a homomorphism into some abelian group. Then f factors through a homomorphism $B/A \to X$ if and only if the kernel of f contains A. That f factors means that there is a commutative diagram

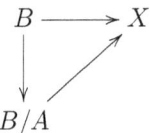

Let $f : A \to B$ be a homomorphism of abelian groups. Then the image $f(A)$ is a subgroup of B. If there is no doubt which homomorphism f is considered, we allow the notation
$$B/A := B/f(A).$$

1.1 Lemma. *A commutative diagram*

$$\begin{array}{ccc} A & \longrightarrow & B \\ \downarrow & & \downarrow \\ C & \longrightarrow & D \end{array}$$

induces homomorphisms

$$B/A \longrightarrow D/C, \quad C/A \longrightarrow D/B.$$

A (finite or infinite) sequence of homomorphisms of abelian groups

$$\cdots \longrightarrow A \longrightarrow B \longrightarrow C \longrightarrow \cdots$$

is called *exact* at B if

$$\text{Kernel}(B \longrightarrow C) = \text{Image}(A \longrightarrow B).$$

It is called exact if it is exact at every place. An exact sequence $A \to B \to C$ induces an injective homomorphism

$$B/A \hookrightarrow C.$$

The sequence $0 \to A \to B$ is exact if and only if $A \to B$ is injective. The sequence $A \to B \to 0$ is exact if and only if $A \to B$ is surjective. The sequence

$$0 \longrightarrow A \longrightarrow B \longrightarrow C \longrightarrow 0$$

is exact if and only if $A \to B$ is injective and if the induced homomorphism $B/A \to C$ is an isomorphism. A sequence of this form is called a *short exact sequence*. Hence the typical short exact sequence is

$$0 \longrightarrow A \longrightarrow B \longrightarrow B/A \longrightarrow 0 \quad (A \subset B).$$

2. Some homological algebra

A *complex* $A^{\bullet}$ is a sequence of homomorphisms of abelian groups (parameterized by $\mathbb{Z}$)

$$\cdots \longrightarrow A_{n-1} \xrightarrow{d_{n-1}} A_n \xrightarrow{d_n} A_{n+1} \longrightarrow \cdots$$

such that the composition of two consecutive is zero, $d_n \circ d_{n-1} = 0$. Usually one omits indices at the d-s and writes simply $d = d_n$ and hence $d \circ d = 0$, which sometimes is written as $d^2 = 0$. The cohomology groups of $A^{\bullet}$ are defined as

$$H^n(A^{\bullet}) := \frac{\text{Kernel}(A^n \to A^{n+1})}{\text{Image}(A^{n-1} \to A^n)} \quad (n \in \mathbb{Z}).$$

They vanish if and only if the complex is exact. Hence the cohomology groups measure the absence of exactness of a complex.

A homomorphism $f^{\bullet} : A^{\bullet} \to B^{\bullet}$ of complexes is a commutative diagram

$$\begin{array}{ccccccccc} \cdots & \longrightarrow & A^{n-1} & \longrightarrow & A^n & \longrightarrow & A^{n+1} & \longrightarrow & \cdots \\ & & \downarrow f^{n-1} & & \downarrow f^n & & \downarrow f^{n+1} & & \\ \cdots & \longrightarrow & B^{n-1} & \longrightarrow & B^n & \longrightarrow & B^{n+1} & \longrightarrow & \cdots \end{array}$$

§2. Some homological algebra

It is clear how to compose two complex homomorphisms $f^{\bullet} : A^{\bullet} \to B^{\bullet}$, $g^{\bullet} : B^{\bullet} \to C^{\bullet}$ to a complex homomorphism $g^{\bullet} \circ f^{\bullet} : A^{\bullet} \to C^{\bullet}$. A sequence of complex homomorphisms

$$\cdots \longrightarrow A^{\bullet} \longrightarrow B^{\bullet} \longrightarrow C^{\bullet} \longrightarrow \cdots$$

is called exact if all the induced sequences

$$\cdots \longrightarrow A^n \longrightarrow B^n \longrightarrow C^n \longrightarrow \cdots$$

are exact. There is also the notion of a short exact sequence of complexes

$$0 \longrightarrow A^{\bullet} \longrightarrow B^{\bullet} \longrightarrow C^{\bullet} \longrightarrow 0.$$

Here 0 stands for the zero complex ($0^n = 0$, $d^n = 0$ for all n).

A homomorphism of complexes $A^{\bullet} \to B^{\bullet}$ induces natural homomorphisms

$$H^n(A^{\bullet}) \longrightarrow H^n(B^{\bullet})$$

of the cohomology groups (use Lemma 1.1). These homomorphisms are compatible with the composition of complex-homomorphisms. A less obvious construction is as follows. Let

$$0 \longrightarrow A^{\bullet} \longrightarrow B^{\bullet} \longrightarrow C^{\bullet} \longrightarrow 0$$

be a short exact sequence of complexes. We construct a homomorphism

$$\delta : H^n(C^{\bullet}) \longrightarrow H^{n+1}(A^{\bullet}).$$

Let $[c] \in H^n(C^{\bullet})$ be represented by an element $c \in C^n$. Take a pre-image $b \in B^n$ and consider $\beta = db \in B^{n+1}$. Since β goes to $d(c) = 0$ in C^{n+1} there exists a pre-image $a \in A^{n+1}$. This goes to 0 in A^{n+2} (because A^{n+2} is imbedded in B^{n+2} and b goes to $d^2(b) = 0$ there). Hence a defines a cohomology class $[a] \in H^{n+1}(A^{\bullet})$. It is easy to check that this class doesn't depend on the above choices.

2.1 Fundamental lemma of homological algebra. Let

$$0 \longrightarrow A^{\bullet} \longrightarrow B^{\bullet} \longrightarrow C^{\bullet} \longrightarrow 0$$

be a short exact sequence of complexes. Then the long sequence

$$\cdots \to H^{n-1}(C^{\bullet}) \xrightarrow{\delta} H^n(A^{\bullet}) \to H^n(B^{\bullet}) \to H^n(C^{\bullet}) \xrightarrow{\delta} H^{n+1}(C^{\bullet}) \to \cdots$$

is exact.

We leave the details to the reader. □

There is a second lemma of homological algebra which we will need.

2.2 Lemma. *Let*

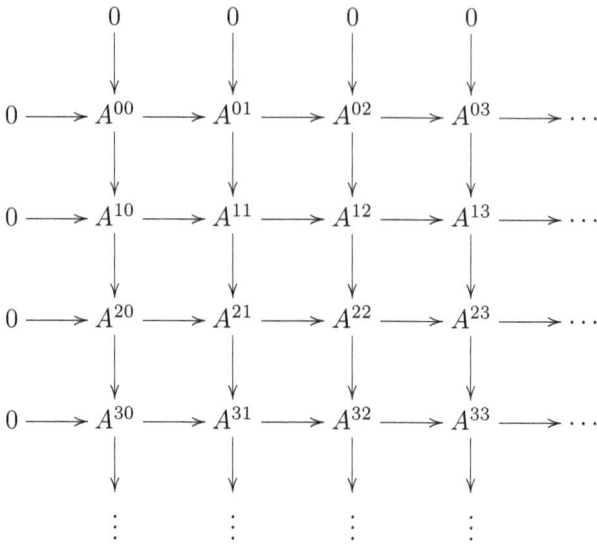

be a commutative diagram where all lines and columns are exact besides the first column and the first row (those containing A^{00}). Then there is a natural isomorphism between the cohomology groups of the first row and the first column,

$$H^n(A^{\cdot,0}) \cong H^n(A^{0,\cdot})$$

For $n = 0$ this is understood as

$$\text{Kernel}(A^{00} \longrightarrow A^{01}) = \text{Kernel}(A^{00} \longrightarrow A^{10}).$$

The proof is given by "diagram chasing". We only give a hint how it works. Assume $n = 1$. Let $[a] \in H^1(A^{0,\cdot})$ be a cohomology class represented by an element $a \in A^{0,1}$. This element goes to 0 in $A^{0,2}$. As a consequence the image of a in $A^{1,1}$ goes to 0 in $A^{1,2}$. Hence this image comes from an element $\alpha \in A^{1,0}$. Clearly, this element goes to zero in $A^{2,0}$ (since it goes to 0 in $A^{2,1}$.) Now α defines a cohomology class $[\alpha] \in H^1(A^{\cdot,0})$. There is some extra work to show that this map is well-defined. □

3. The tensor product

All rings which we consider are assumed to be commutative and with unit element. Ring homomorphisms are assumed to map the unit element into the unit element. A module M over a ring A is an abelian group together with a map $A \times M \to M$, $(a, m) \mapsto am$, such that the usual axioms of a vector space are satisfied including $1_A m = m$ for all $m \in M$. The notion of linear maps, kernel, image of a linear map are as in the case of vector spaces. But in contrast to the case of vector spaces, a module has usually no basis. A module which admits a basis is called free. A finitely generated free module is isomorphic to R^n.

If $M \subset N$ is a submodule, then the factor group N/M carries a structure of an A-module. All what we have said about exact sequences of abelian groups is literally true for A-modules.

Tensor product

Recall that for two modules M, N over a ring R, there exists a module $M \otimes_R N$ together with an R-bilinear map

$$M \times N \longrightarrow M \otimes_R N, \quad (a, b) \mapsto a \otimes b,$$

such that for each bilinear map $M \times N \to P$ into an arbitrary third module P there exists a unique commutative diagram

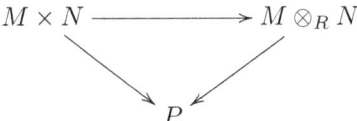

with an R-linear map $M \otimes_R N \to P$. The tensor product $M \otimes_R N$ is generated by the special elements $m \otimes n$.

If $f : M \to M'$ and $g : N \to N'$ are R-linear maps, then one gets a natural R-linear map

$$f \otimes g : M \otimes_R N \longrightarrow M' \otimes_R N', \quad a \otimes b \mapsto f(a) \otimes g(a).$$

It is clear that this map is uniquely determined by this formula. The existence follows from the universal property applied to the map $(a, b) \mapsto f(a) \otimes g(b)$.

Basic properties of the tensor product

There is a natural isomorphism
$$R \otimes_R M \xrightarrow{\sim} M, \quad r \otimes m \mapsto rm,$$
and more generally
$$R^n \otimes_R M \xrightarrow{\sim} M^n.$$
As a special case we get
$$R^n \otimes_R R^m \cong R^{n \times m}.$$
This is related also to the formula
$$(M \times N) \otimes_R P \cong (M \otimes_R P) \times (N \otimes_R P) \quad \text{(canonically)}.$$
The tensor product is associative: for R-modules M, N, P one has an isomorphism
$$(M \otimes_R N) \otimes_R P \xrightarrow{\sim} M \otimes_R (N \otimes_R P), \quad (m \otimes n) \otimes p \mapsto m \otimes (n \otimes p).$$
The existence of this map follows from the universal property of the tensor product.

The tensor product is also commutative in the following sense:
$$M \otimes_R N \xrightarrow{\sim} N \otimes_R M, \quad m \otimes n \mapsto n \otimes m.$$

Ring extension

Let $A \to B$ be a ring homomorphism and M an A-module. Then $M \otimes_A B$ carries a natural structure as B-module. It is given by $b(m \otimes b') = m \otimes (bb')$. The existence follows from the universal property of the tensor product. A special case is
$$A^n \otimes_A B \cong B^n.$$

Existence of the tensor product

For an arbitrary set I we define R^I to be the set of all maps $I \to R$, $i \mapsto r_i$, such that r_i is 0 for almost all i. So $R^I = R^n$ for $I = \{1, \ldots, n\}$. By definition a module is free if and only if it is isomorphic to an R^I for suitable I. An arbitrary R-module M can be represented by an exact sequence
$$R^J \longrightarrow R^I \longrightarrow M \longrightarrow 0.$$
For another R-module N we define now the tensor product by the exact sequence
$$N^J \longrightarrow N^I \longrightarrow M \otimes_R N \longrightarrow 0.$$
The bilinear map $M \times N \to M \otimes_R N$ and the universal property are obvious. In particular, this construction is independent of the choice of the representation (up to canonical isomorphism).

§3. The tensor product

Exactness properties

Let $M \to N$ be an injective homomorphism of R-modules. For an R-module P the induced homomorphism $M \otimes_R P \to N \otimes_R P$ needs not to be injective. But when $P \cong R^n$ is free, injectivity is preserved. A slight and trivial extension of this observation is the following remark.

3.1 Remark. *Let $M_1 \to M_2 \to M_3$ be an exact exact sequence of R-modules. Then for every **free** module P the sequence $M_1 \otimes_R P \to M_2 \otimes_R P \to M_3 \otimes_R P$ remains exact.*

In this context we mention some other exactness properties. For two R-modules M, N we denote by $\mathrm{Hom}_R(M,M)$ the set of all R-linear maps $M \to N$. This is an R-module. Let $M \to N$ be an R-linear map. Then for an arbitrary R-module P one has obvious R-linear maps

$$\mathrm{Hom}_R(P, M) \longrightarrow \mathrm{Hom}_R(P, N), \qquad \mathrm{Hom}_R(N, P) \longrightarrow \mathrm{Hom}_R(M, P).$$

Since $\mathrm{Hom}(R^n, M) \cong M^n$, one obtains the following result.

3.2 Remark. *Let $0 \to M_1 \to M_2 \to M_3 \to 0$ be an exact sequence of R-modules and P also an R-module, then the following holds.*
a) *If P is free then*

$$0 \longrightarrow \mathrm{Hom}_R(P, M_1) \longrightarrow \mathrm{Hom}_R(P, M_2) \longrightarrow \mathrm{Hom}_R(P, M_3) \longrightarrow 0$$

remains exact.
b) *If M_3 is free then*

$$0 \longrightarrow \mathrm{Hom}_R(M_3, P) \longrightarrow \mathrm{Hom}_R(M_2, P) \longrightarrow \mathrm{Hom}_R(M_1, P) \longrightarrow 0$$

remains exact.

We comment shortly on b). When M_3 is free, one can choose a system of elements in M_2 whose images in M_3 define a basis. This system generates a submodule $M_3' \subset M_2$ which maps isomorphically to M_3. Now it is easy to see that M_2 is isomorphic to $M_1 \times M_3$ and the map $M_1 \to M_2$ corresponds to $m \mapsto (m, 0)$ and the map $M_2 \to M_3$ corresponds to $(m_1, m_3) \mapsto m_3$. Now the exactness should be clear. □

Chapter III. Sheaves

1. Presheaves

We introduce the language of presheaves of abelian groups. This consists mainly of definitions and simple remarks whose proofs are very simple. In many cases they can be left to the reader.

1.1 Definition. *A presheaf F (of abelian groups) on a topological space X is a map which assigns to every open subset $U \subset X$ an abelian group $F(U)$ and to every pair U, V of open subsets with the property $V \subset U$ a homomorphism*
$$r_V^U : F(U) \longrightarrow F(V)$$
such that $r_U^U = \mathrm{id}$ and such that for three open subsets U, V, W with the property $W \subset V \subset U$ the relation
$$r_W^U = r_W^V \circ r_V^U$$
holds.

Example: $F(U)$ is the set of all continuous functions $f : U \to \mathbb{C}$ and $r_V^U(f) := f|V$ (restriction).

Many presheaves generalize this example. Hence the maps r_V^U are called "restrictions" in general and one uses the notation
$$s|V = s|_F V := r_V^U(s) \qquad \text{for} \quad s \in F(U).$$
The elements of $F(U)$ sometimes are called "sections" of F over U. In the special case $U = X$ they are called "global" sections.

1.2 Definition. *Let X be a topological space. A homomorphism of presheaves*
$$f : F \longrightarrow G$$
is a family of group homomorphisms
$$f_U : F(U) \longrightarrow G(U),$$
such that the diagram
$$\begin{array}{ccc} F(U) & \longrightarrow & G(U) \\ \downarrow & & \downarrow \\ F(V) & \longrightarrow & G(V) \end{array}$$
commutes for every pair $V \subset U$ of open subsets, i.e. $f_U(s)|_G V = f_V(s|_F V)$.

§2. Germs and Stalks

It is clear how to define the identity map $\mathrm{id}_F : F \to F$ of a presheaf and the composition $g \circ f$ of two homomorphisms $f : F \to G$, $g : G \to H$ of presheaves.

There is also a natural notion of a sub-presheaf $F \subset G$. Besides $F(U) \subset G(U)$ for all U, one has to demand that the restrictions are compatible. This means:

The canonical inclusions $i_U : F(U) \to G(U)$ define a homomorphism $i : F \to G$ of presheaves.

When $f : F \to G$ is a homomorphism of presheaves, the images $f_U(F(U))$ define a sub-presheaf of G. We call it the *presheaf-image* and denote it by

$$f_{\mathrm{pre}}(F).$$

It is also clear that the kernels of the maps f_U define a sub-presheaf of F. We denote it by $\mathrm{Kernel}(f : F \to G)$. When F is a sub-presheaf of G, then one can consider the factor groups $G(U)/F(U)$. Using Lemma II.1.1 it is clear how to define restriction maps to get a presheaf $G/_{\mathrm{pre}}F$. We call this presheaf the *factor presheaf*.

Since we have defined kernel and image, we can also introduce the notion of a *presheaf-exact sequence*. A sequence $F \to G \to H$ is presheaf-exact if and only if $F(U) \to G(U) \to H(U)$ is exact for all U. What we have said about exact sequences of abelian groups carries literally over to presheaf-exact sequences of presheaves of abelian groups.

2. Germs and Stalks

Let F be a presheaf on a topological space X and let $a \in X$ be a point. We consider pairs (U, s), where U is an open neighbourhood of a and where $s \in F(U)$ is a section over U. Two pairs (U, s), (V, t) are called equivalent if there exists an open neighborhood $a \in W \subset U \cap V$, such that $s|W = t|W$. This is an equivalence relation. The equivalence classes

$$[U, s]_a := \{ (V, t);\quad (V, t) \sim (U, s) \}$$

are called *germs* of F at the point a. The set of all germs

$$F_a := \{ [U, s]_a,\quad a \in U \subset X,\ s \in F(U) \}$$

is the so-called *stalk* of F at a. The stalk carries a natural structure as abelian group. One defines

$$[U, s]_a + [V, t]_a := [U \cap V, s|U \cap V + t|U \cap V]_a.$$

We use sometimes the simplified notation
$$s_a = [U, s]_a.$$
For every open neighborhood $a \in U \subset X$ there is an obvious homomorphism
$$F(U) \longrightarrow F_a, \quad s \longmapsto s_a.$$
A homomorphism of presheaves $f : F \to G$ induces natural mappings
$$f_a : F_a \longrightarrow G_a \quad (a \in X).$$
The image of a germ $[U, s]_a$ is simply $[U, f_U(s)]_a$. It is easy to see that this is well-defined.

2.1 Remark. *Let $F \to G$ and $G \to H$ be homomorphisms of presheaves and let $a \in X$ be a point. Assume that every point a contains arbitrarily small open neighborhoods U such that $F(U) \to G(U) \to H(U)$ is exact. Then $F_a \to G_a \to H_a$ is exact.*

Corollary. *If $F \to G \to H$ is presheaf-exact then $F_a \to G_a \to H_a$ is exact for all a.*

("Arbitrarily small" means that each neighborhood W of a contains a U.) The proof is easy and can be omitted. □

For a sub-presheaf $F \subset G$ the natural homomorphisms $F_a \to G_a$ are injective. Usually we will identify F_a with its image in G_a. In particular, for a homomorphism $F \to G$ of presheaves and a point $a \in X$, $f_a(F_a)$ and $f_{\text{pre}}(F)_a$ both are subgroups of G_a. It is easy to check that they are equal.
$$f_{\text{pre}}(F)_a = f_a(F_a).$$
If F is a presheaf on X, one can consider for each open subset $U \subset X$
$$F^{(0)}(U) := \prod_{a \in U} F_a.$$
The elements are families $(s_a)_{a \in U}$ with $s_a \in F_a$. There is no coupling between the different s_a. Hence $F^{(0)}(U)$ usually is very monstrous.

For open sets $V \subset U$, one has an obvious homomorphism (projection) $F^{(0)}(U) \to F^{(0)}(V)$. Hence we obtain a presheaf $F^{(0)}$ together with a natural homomorphism $F \longrightarrow F^{(0)}$. Each homomorphism $F \to G$ of presheaves induces a homomorphism $F^{(0)} \to G^{(0)}$ such that the diagram
$$\begin{array}{ccc} F & \longrightarrow & G \\ \downarrow & & \downarrow \\ F^{(0)} & \longrightarrow & G^{(0)} \end{array}$$
commutes.

3. Sheaves

3.1 Definition. *A presheaf F is called a **sheaf** if the following conditions are satisfied:*

(G1) *When $U = \bigcup U_i$ is an open covering of an open subset $U \subset X$ and if $s, t \in F(U)$ are sections with the property $s|U_i = t|U_i$ for all i, then $s = t$.*

(G2) *When $U = \bigcup U_i$ is an open covering of an open subset $U \subset X$ and if $s_i \in F(U_i)$ is a family of sections with the property*
$$s_i|U_i \cap U_j = s_j|U_i \cap U_j \quad \text{for all } i, j,$$
then there exists a section $s \in F(U)$ with the property $s|U_i = s_i$ for all i.

(G3) *$F(\emptyset)$ is the zero group.*

The presheaf of continuous functions clearly is a sheaf, since continuity is a local property. An example of a presheaf F, which usually is not a sheaf, is the presheaf of constant functions with values in $\mathbb{Z}$ ($F(U) = \{f : U \to \mathbb{Z},\ f$ constant$\}$). But the set of *locally constant* functions with values in $\mathbb{Z}$ is a sheaf.

By a subsheaf of a sheaf F we understand a sub-presheaf $G \subset F$ which is already a sheaf. If F, G are sheaves, then a homomorphism $f : F \to G$ of presheaves is called also a homomorphism of sheaves.

3.2 Remark. *Let $F \subset G$ be a sub-presheaf. We assume that G (but not necessarily F) is a sheaf. Then there is a smallest subsheaf $\tilde{F} \subset G$ which contains F. For an arbitrary point $a \in X$ the induced map $f_a : F_a \to \tilde{F}_a$ is an isomorphism.*

Proof. It is clear that $\tilde{F}(U)$ has to be defined as set of all $s \in G(U)$ such that there exists an open covering $U = \bigcup U_i$, such that $s|U_i$ is in $F(U_i)$ for all i. This is equivalent with: the germ s_a is in F_a for all $a \in U$, i.e.
$$\tilde{F}(U) = \{s \in G(U);\ s_a \in F_a \text{ for } a \in U\}. \qquad \square$$

We mention the trivial fact that a section $s \in F(U)$ of a sheaf is zero if all its germs $s_a \in F_a$ are zero for $a \in U$.

3.3 Definition. *Let $F \to G$ be a homomorphism of sheaves. The sheaf-image $f_{\text{sheaf}}(F)$ is the smallest subsheaf of G, which contains the presheaf-image $f_{\text{pre}}(F)$.*

We mentioned above the formula $f_{\text{pre}}(F)_a = f_a(F_a)$. Applying Remark 3.2 gives the formula $f_{\text{sheaf}}(F)_a = f_a(F_a)$ for a homomorphism $F \to G$ of sheaves.

We have to differ between two natural notions of surjectivity.

3.4 Definition.
1) *A homomorphism of presheaves $f : F \to G$ is called **presheaf-surjective** if $f_{\mathrm{pre}}(F) = G$.*

2) *A homomorphism of sheaves $f : F \to G$ is called **sheaf-surjective** if $f_{\mathrm{sheaf}}(F) = G$.*

When F and G both are sheaves, then sheaf-surjectivity and presheaf-surjectivity are different things. We give an example which will be basic.

Let $\mathcal{O}$ be the sheaf of holomorphic functions on $\mathbb{C}$, hence $\mathcal{O}(U)$ is the set of all holomorphic functions on an open subset U. This is a sheaf of abelian groups (under addition). Similarly, we consider the sheaf $\mathcal{O}^*$ of holomorphic functions without zeros. This is also a sheaf of abelian groups (under multiplication). The map $f \to e^f$ defines a sheaf homomorphism

$$\exp : \mathcal{O} \longrightarrow \mathcal{O}^*.$$

The map $\mathcal{O}(U) \to \mathcal{O}^*(U)$ is not always surjective. For example for $U = \mathbb{C}^{\bullet}$ the function $1/z$ is not in the image. Hence exp is not presheaf-surjective. But it is know from complex calculus that $\exp : \mathcal{O}(U) \to \mathcal{O}^*(U)$ is surjective if U is simply connected, for example for a disk U. Since a point admits arbitrarily small neighborhoods which are disks, it follows that exp is sheaf-surjective.

3.5 Remark. *A homomorphism of sheaves $f : F \to G$ is sheaf-surjective if and only if the maps $f_a : F_a \to G_a$ are surjective for all $a \in X$.*

We omit the simple proof. □

Fortunately, the notion "injective" does not contain this difficulty. For trivial reason the following remark is true.

3.6 Remark. *Let $f : F \to G$ be a homomorphism of sheaves. The kernel in the sense of presheaves is already a sheaf.*

Hence we don't have to distinguish between presheaf-injective and sheaf-injective and also not between presheaf-kernel and sheaf-kernel.

3.7 Remark. *A homomorphism of sheaves $f : F \to G$ is injective if and only if the maps $f_a : F_a \to G_a$ are injective for all $a \in X$.*

A homomorphism of (pre)sheaves $f : F \to G$ is called an isomorphism if all $F(U) \to G(U)$ are isomorphisms. Their inverses then define a homomorphism $f^{-1} : G \to F$.

3.8 Remark. *A homomorphism of sheafs $F \to G$ is an isomorphism if and only if $F_a \to G_a$ is an isomorphism for all a.*

For presheaves this is false. As counter example one can take for F the presheaf of constant functions and for G the sheaf of locally constant functions.

It is natural to introduce the notion of sheaf-exactness as follows:

3.9 Definition. *A sequence $F \to G \to H$ of sheaf homomorphisms is **sheaf-exact** at G if the kernel of $G \to H$ and the sheaf-image of $F \to G$ agree.*

Generalizing the remarks 3.5 and 3.7 one can easily show the following proposition.

3.10 Proposition. *A sequence $F \to G \to H$ is exact if and only if $F_a \to G_a \to H_a$ is exact for all a.*

We indicate the proof. We make use of the mentioned formula $f_{\text{sheaf}}(F)_a = f_a(F_a)$. This shows that we can replace F by its sheaf image in G. Hence we can assume that F is a subsheaf of G and $F \to G$ is the natural injection. We have to show that the exactness of the sequences $F_a \to G_a \to H_a$ implies that F is the kernel of $G \to H$. It is clear that F is contained in the kernel. Hence it suffices to show the following. Let $s \in G(U)$ be an element of the kernel $G(U) \to H(U)$. Then we know that the germs s_a are in the kernel of $G_a \to H_a$. Hence they are contained in F_a. This means that there exists an open covering $U = \bigcup U_i$ such that $s|U_i \in F(U_i)$. The sheaf axiom G2 implies that they glue to an element of $F(U)$. The sheaf axiom G1 then shows that this element agrees with s. □

Our discussion so far has obviously one gap. Let $F \subset G$ be a subsheaf of a sheaf G. We would like to have an exact sequence

$$0 \longrightarrow F \longrightarrow G \longrightarrow H \longrightarrow 0.$$

The sheaf H should be the factor sheaf of G by F. But up to now we only defined the factor presheaf $G/_{\text{pre}}F$ which usually is not a sheaf. In the next section we will give the correct definition for a factor sheaf $G/_{\text{sheaf}}F$.

4. The generated sheaf

For a presheaf F we introduced the monstrous presheaf

$$F^{(0)}(U) = \prod_{a \in U} F_a.$$

Obviously $F^{(0)}$ is a sheaf. Sometimes it is called the "Godement sheaf" or the "associated flabby sheaf". There is a natural homomorphism

$$F \longrightarrow F^{(0)}.$$

We can consider its presheaf-image and then the smallest subsheaf which contains it. We denote this sheaf by $\hat{F}$ and call it the "generated sheaf" by F. There is a natural homomorphism

$$F \longrightarrow \hat{F}.$$

From the construction follows immediately

4.1 Remark. *Let F be a presheaf. The natural maps*

$$F_a \xrightarrow{\sim} \hat{F}_a$$

are isomorphisms.

A homomorphism $F \to G$ of presheaves induces a homomorphism $F^{(0)} \to G^{(0)}$. Clearly $\hat{F}$ is mapped into $\hat{G}$. This gives us the following result.

4.2 Remark. *Let $f : F \to G$ be a homomorphism of presheaves. There is a natural homomorphism $\hat{F} \to \hat{G}$, such that the diagram*

$$\begin{array}{ccc} F & \longrightarrow & G \\ \downarrow & & \downarrow \\ \hat{F} & \longrightarrow & \hat{G} \end{array}$$

commutes.

When F is already a sheaf, then $F \to F^{(0)}$ is injective. Then the map of F into the presheaf-image is an isomorphism. This implies that the presheaf-image is already a sheaf.

4.3 Remark. *Let F be a sheaf. Then $F \to \hat{F}$ is an isomorphism.*

If F is a sub-presheaf of a sheaf G, then the induced map $\hat{F} \to \hat{G} \cong G$ is an isomorphism $\hat{F} \to \tilde{F}$ between $\hat{F}$ and the smallest subsheaf $\tilde{F}$ of G, wich contains F.

Hence we can identify $\tilde{F}$ and $\hat{F}$.

Factor sheaves and exact sequences of sheaves

Let $F \to G$ be a homomorphism of presheaves. We introduced already the factor presheaf $G/_{\text{pre}}F$ which associates to an open U the factor group $G(U)/F(U)$. Even if both F and G are sheaves this will usually be not a sheaf. Hence we define the factor sheaf as the sheaf generated by the factor presheaf.

$$G/_{\text{sheaf}}F := \widehat{G/_{\text{pre}}F}.$$

Since we are interested mainly in sheaves, we will write usually for a homomorphism of sheaves $f : F \to G$:

$$G/F := G/_{\text{sheaf}}F \quad \text{(factor sheaf)}$$
$$f(F) := f_{\text{sheaf}}(F) \quad \text{(sheaf image)}$$

Notice that there is no need to differ between sheaf- and presheaf-kernel. When we talk about an exact sequence of sheaves

$$F \longrightarrow G \longrightarrow H,$$

§4. The generated sheaf

we usually mean "sheaf exactness". All what we have said about exactness properties of sequences of abelian groups is literally true for sequences of sheaves. For example: a sequence of sheaves $0 \to F \to G$ (0 denotes the zero sheaf) is exact if and only if $F \to G$ is injective. A sequence of sheaves $F \to G \to 0$ is exact if and only if $F \to G$ is surjective (in the sense of sheaves of course). A sequence of sheaves $0 \to F \to G \to H \to 0$ is exact if and only if there is an ismomorphism $H \cong G/F$ which identifies this sequence with

$$0 \longrightarrow F \longrightarrow G \longrightarrow G/F \longrightarrow 0.$$

4.4 Remark. Let $0 \to F \to G \to H \to 0$ be an exact sequence of sheaves. Then for open U the sequence

$$0 \to F(U) \to G(U) \to H(U)$$

is exact.

Corollary. *The sequence*

$$0 \to F(X) \to G(X) \to H(X)$$

is exact.

The simple proof can be left to the reader. □

Usually $G(X) \to H(X)$ is not surjective as the example

$$0 \longrightarrow \mathbb{Z}_X \longrightarrow \mathcal{O} \xrightarrow{f \mapsto e^{2\pi i f}} \mathcal{O}^* \longrightarrow 0$$

shows. Here $\mathbb{Z}_X$ denotes the sheaf of locally constant functions with values in $\mathbb{Z}$. Cohomology theory will measure the absence of right exactness. The above sequence will be part of a long exact sequence

$$0 \longrightarrow F(X) \longrightarrow G(X) \longrightarrow H(X) \longrightarrow H^1(X, F) \longrightarrow \cdots$$

Chapter IV. Cohomology of sheaves

1. The canonical flabby resolution

A sheaf F is called *flabby* if $F(X) \to F(U)$ is surjective for all open U. Then $F(U) \to F(V)$ is surjective for all $V \subset U$. An example for a flabby sheaf is the Godement sheaf $F^{(0)}$. Recall that we have the exact sequence

$$0 \longrightarrow F \longrightarrow F^{(0)}.$$

We want to extend this sequence. For this we consider the sheaf $F^{(0)}/F$ and embed it into its Godement sheaf,

$$F^{(1)} := \left(F^{(0)}/F\right)^{(0)}.$$

In this way we get a long exact sequence

$$0 \longrightarrow F \longrightarrow F^{(0)} \longrightarrow F^{(1)} \longrightarrow F^{(2)} \longrightarrow \cdots$$

If $F^{(n)}$ has been already constructed, then we define

$$F^{(n+1)} := \left(F^{(n)}/F^{(n-1)}\right)^{(0)}.$$

The sheaves $F^{(n)}$ are all flabby. We call this sequence the *canonical flabby resolution* or the *Godement resolution*. Sometimes it is useful to write the resolution in the form

$$\begin{array}{ccccccccccc}
\cdots \longrightarrow & 0 & \longrightarrow & F & \longrightarrow & 0 & \longrightarrow & 0 & \longrightarrow & 0 & \longrightarrow \cdots \\
& \downarrow & & \downarrow & & \downarrow & & \downarrow & & \downarrow & \\
\cdots \longrightarrow & 0 & \longrightarrow & F^{(0)} & \longrightarrow & F^{(1)} & \longrightarrow & F^{(2)} & \longrightarrow & F^{(3)} & \longrightarrow \cdots
\end{array}$$

Both lines are complexes. The vertical arrows can be considered as a complex homomorphism. The induced homomorphism of the cohomology groups are isomorphisms. Notice that only the 0-cohomology group of both complexes is different from 0. This zero cohomology group is naturally isomorphic F.

Now we apply the global section functor Γ to the resolution. This is

$$\Gamma F := F(X).$$

§1. The canonical flabby resolution

We obtain a long sequence

$$0 \longrightarrow \Gamma F \longrightarrow \Gamma F^{(0)} \longrightarrow \Gamma F^{(1)} \longrightarrow \Gamma F^{(2)} \longrightarrow \cdots$$

The essential point is that this sequence is no longer exact. We only can say that it is a complex. We prefer to write it in the form

$$\begin{array}{ccccccccccc}
\cdots & \longrightarrow & 0 & \longrightarrow & \Gamma F & \longrightarrow & 0 & \longrightarrow & 0 & \longrightarrow & 0 & \longrightarrow \cdots \\
& & \downarrow & & \downarrow & & \downarrow & & \downarrow & & \downarrow & \\
\cdots & \longrightarrow & 0 & \longrightarrow & \Gamma F^{(0)} & \longrightarrow & \Gamma F^{(1)} & \longrightarrow & \Gamma F^{(2)} & \longrightarrow & \Gamma F^{(3)} & \longrightarrow \cdots
\end{array}$$

The second line is

$$\cdots \longrightarrow 0 \longrightarrow \Gamma F^{(0)} \longrightarrow \Gamma F^{(1)} \longrightarrow \Gamma F^{(2)} \longrightarrow \cdots$$
$$\uparrow$$
$$\text{zero position}$$

Now we define the cohomology groups $H^{\cdot}(X, F)$ to be the cohomology groups of this complex:

$$H^n(X, F) := \frac{\operatorname{Kern}(\Gamma F^{(n)} \longrightarrow \Gamma F^{(n+1)})}{\operatorname{Image}(\Gamma F^{(n-1)} \longrightarrow \Gamma F^{(n)})}.$$

(We define $\Gamma F^{(n)} = 0$ for $n < 0$.) Clearly

$$H^n(X, F) = 0 \quad \text{for} \quad n < 0.$$

Next we treat the special case $n = 0$,

$$H^0(X, F) = \operatorname{Kernel}(\Gamma F^{(0)} \longrightarrow \Gamma F^{(1)}).$$

Since the kernel can be taken in the presheaf sense, we can write

$$H^0(X, F) = \Gamma \operatorname{Kernel}(F^{(0)} \longrightarrow F^{(1)}).$$

Recall that $F^{(1)}$ is a sheaf which contains $F^{(0)}/F$ as subsheaf. We obtain

$$H^0(X, F) = \Gamma \operatorname{Kernel}(F^{(0)} \longrightarrow F^{(0)}/F)$$

This is the image of F in $F^{(0)}$ and hence a sheaf which is canonically isomorphic to F.

36 Chapter IV. Cohomology of sheaves

1.1 Remark. *There is a natural isomorphism*
$$H^0(X, F) \cong \Gamma F = F(X).$$

If $F \to G$ is a homomorphism of sheaves, then the homomorphism $F_a \to G_a$ induces a homomorphism $F^{(0)} \to G^{(0)}$. More generally, one has an obvious natural homomorphism $F^{(n)} \to G^{(n)}$ for all n. This gives a homomorphism of the Godement resolution. Hence we obtain a natural homomorphism
$$H^n(X, F) \longrightarrow H^n(X, G).$$

If $F \to G \to H$ is an exact sequence, then $F^{(0)} \to G^{(0)} \to H^{(0)}$ is also exact (already as sequence of presheaves). More generally, the following lemma holds.

1.2 Lemma. *Let $F \to G \to H$ be an exact sequence of sheaves. Then the induced sequence $F^{(n)} \to G^{(n)} \to H^{(n)}$ is exact for every n.*

Proof. By a general principle it is sufficient to prove that $F \mapsto F^{(n)}$ maps short exact sequences $0 \to F \to G \to H \to 0$ into short exact sequences $0 \to F^{(n)} \to G^{(n)} \to H^{(n)} \to 0$. The reason is that an arbitray exact sequence $F \xrightarrow{f} G \xrightarrow{g} H$ can be splitted into short exact sequences

$$0 \longrightarrow \mathrm{Kernel}(f) \longrightarrow F \longrightarrow f(F) \longrightarrow 0,$$
$$0 \longrightarrow f(F) \longrightarrow G \longrightarrow g(G) \longrightarrow 0,$$
$$0 \longrightarrow g(G) \longrightarrow H \longrightarrow H/g(G) \longrightarrow 0.$$

So we start with a short exact sequence $0 \to F \to G \to H \to 0$. The proof can now be given by induction. One needs the following lemma about abelian groups:

Let

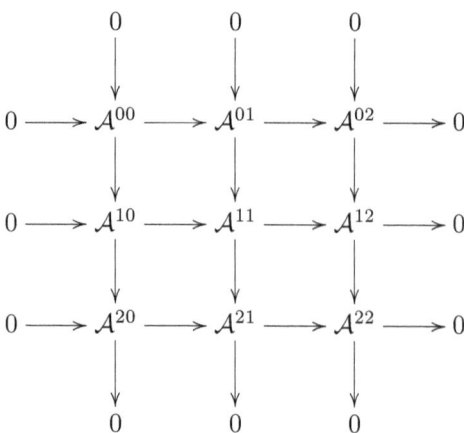

§1. The canonical flabby resolution

be a commutative diagram such that the three columns and the first two lines are exact. Then the third line is also exact.

The proof is easy and can be omitted. □

Before we continue, we need a basic lemma:

1.3 Lemma. Let $0 \to F \to G \to H \to 0$ be a short exact sequence of sheaves. Assume that F is flabby. Then

$$0 \longrightarrow \Gamma F \longrightarrow \Gamma G \longrightarrow \Gamma H \longrightarrow 0$$

is exact.

Proof. Let $h \in H(X)$. We have to show that h is the image of an $g \in G(X)$. For the proof one considers the set of all pairs (U, g), where U is an open subset and $g \in G(U)$ and such that g maps to $h|U$. This set is ordered by

$$(U, g) \geq (U', g') \iff U' \subset U \text{ and } g|U' = g'.$$

From the sheaf axioms follows that every inductive subset has an upper bound. (Take the union of all open subsets which occur in the inductive set.) By Zorns's lemma there exists a maximal (U, g). We have to show $U = X$. If this is not the case, we can find a pair (U', g') in the above set such that U' is not contained in U. The difference $g - g'$ defines a section in $F(U \cap U')$. Since F is flabby, this extends to a global section. This allows us to modify g' such that it glues with g to a section on $U \cup U'$. □

An immediate corollary of Lemma 1.3 states:

1.4 Lemma. Let $0 \to F \to G \to H \to 0$ an exact sequence of sheaves. If F and G are flabby then H is flabby too.

Let $0 \to F \to G \to H \to 0$ be an exact sequence of sheafs. We obtain a commutative diagram

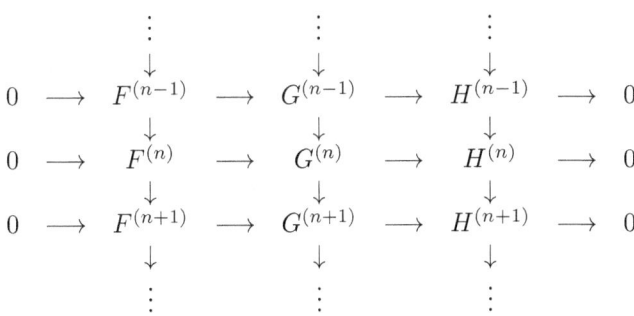

From Lemma 1.2 we know that all lines of this diagram are exact. From Lemma 1.3 follows that they remain exact after applying Γ. Hence the diagram

$$
\begin{array}{ccccccccc}
& & \vdots & & \vdots & & \vdots & & \\
& & \downarrow & & \downarrow & & \downarrow & & \\
0 & \to & \Gamma F^{(n-1)} & \to & \Gamma G^{(n-1)} & \to & \Gamma H^{(n-1)} & \to & 0 \\
& & \downarrow & & \downarrow & & \downarrow & & \\
0 & \to & \Gamma F^{(n)} & \to & \Gamma G^{(n)} & \to & \Gamma H^{(n)} & \to & 0 \\
& & \downarrow & & \downarrow & & \downarrow & & \\
0 & \to & \Gamma F^{(n+1)} & \to & \Gamma G^{(n+1)} & \to & \Gamma H^{(n+1)} & \to & 0 \\
& & \downarrow & & \downarrow & & \downarrow & & \\
& & \vdots & & \vdots & & \vdots & &
\end{array}
$$

can be considered as a short exact sequence of complexes. We can apply Lemma II.2.1 to obtain the long exact cohomology sequence:

1.5 Theorem. *Every short exact sequence $0 \to F \to G \to H \to 0$ induces a natural long exact cohomology sequence*

$$0 \to \Gamma F \to \Gamma G \to \Gamma H \xrightarrow{\delta} H^1(X, F) \to H^1(X, G) \to H^1(X, H)$$
$$\xrightarrow{\delta} H^2(X, F) \to \cdots$$

The next Lemma shows that the cohomology of flabby sheaves is trivial.

1.6 Lemma. *Let*

$$0 \to F \to F_0 \to F_1 \to \cdots$$

be an exact sequence of flabby sheaves (finite or infinite). Then

$$0 \to \Gamma F \to \Gamma F_0 \to \Gamma F_1 \to \cdots$$

is exact.

Corollary. *For flabby F one has:*

$$H^i(X, F) = 0 \quad \text{for} \quad i > 0.$$

Proof. We use the splitting principle. The long exact sequence can be splitted into short exact sequences

$$0 \to F \to F_0 \to F_0/F \to 0, \quad 0 \to F_0/F \to F_1 \to F_1/F_0 \to 0, \ldots$$

From Lemma 1.4 we get that $F_0/F, F_1/F_0, \ldots$ are flabby. The claim now follows from Lemma 1.3. □

A sheaf F is called *acyclic* if $H^n(X, F) = 0$ for $n > 0$. Hence flabby sheaves are acyclic. By an *acyclic* resolution of a sheaf we understand an exact sequence

$$0 \to F \to F_0 \to F_1 \to F_2 \to \cdots$$

with acyclic F_i.

§2. Sheaves of rings and modules

1.7 Proposition. *Let $0 \to F \to F_0 \to F_1 \to \cdots$ be an acyclic resolution of F. Then there is a natural isomorphism between the n-the cohomology group $H^n(X, F)$ and the n-th cohomology group of the complex*

$$\cdots \longrightarrow 0 \longrightarrow \Gamma F_0 \longrightarrow \Gamma F_1 \longrightarrow \Gamma F_2 \longrightarrow \cdots$$
$$\uparrow$$
$$\textit{zero position}$$

Proof. Taking the canonical flabby resolutions of F and of all F_n on gets a diagram

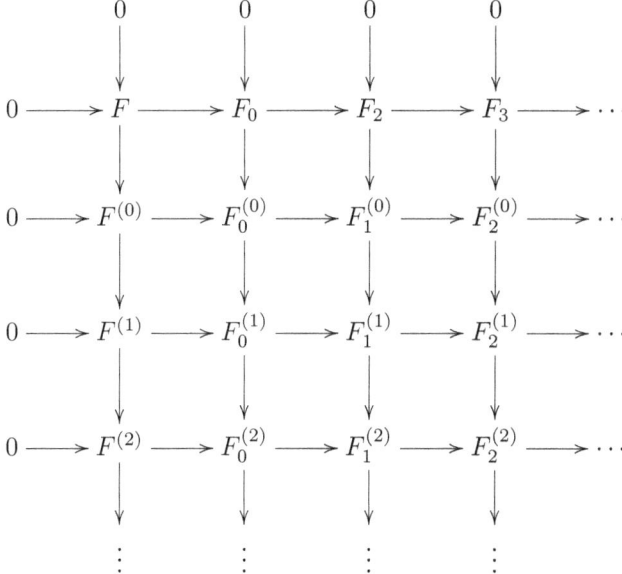

All lines and columns are exact. We apply Γ to this complex. Then all lines and columns besides the first ones remain exact. We can apply Lemma II.2.2. □

2. Sheaves of rings and modules

Let A be a (commutative and unital) ring. A sheaf of A-modules is a sheaf F of abelian groups such that every $F(U)$ carries a structure as A-module and such the the restriction maps $F(U) \to F(V)$ for $V \subset U$ are A-linear. A homomorphism $F \to G$ is called A-linear if all $F(U) \to G(U)$ are so. Then kernel and image carry natural structures of sheafs of A-modules. Also the stalks carry such a structure naturally. Hence the whole canonical flabby resolution is a sequence of sheafs of A-modules. This implies that the cohomology groups also are A-modules.

There is a refinement of this construction: By a sheaf of rings $\mathcal{O}$ we understand a sheaf of abelian groups such that every $\mathcal{O}(U)$ is not only an abelian group but a ring and such that all restriction maps $\mathcal{O}(U) \to \mathcal{O}(V)$ are ring homomorphisms. Then the stalks $\mathcal{O}_a$ carry a natural ring structure such that the homomorphisms $\mathcal{O}(U) \longrightarrow \mathcal{O}_a$ (U is an open neighborhood of a) are ring homomorphisms.

By an $\mathcal{O}$-module we understand a sheaf $\mathcal{M}$ of abelian groups such that every $\mathcal{M}(U)$ carries a structure as $\mathcal{O}(U)$-module and such that the restriction maps are compatible with the module structure. To make this precise we give a short comment. Let M be an A-module and N be a module over a different ring B. Asssume that a homomorphism $r: A \to B$ is given. A homomorphism $f: M \to N$ of abelian groups is called compatible with the module structures if the formula

$$f(am) = r(a)f(m) \qquad (a \in A, \ m \in M)$$

holds. An elegant way to express this is as follows. We can consider N also as an module over A by means of the definition $an := r(a)n$. Sometimes this A-module is written as $N_{[r]}$. Then the compatibility of the map f simply means that it is an A-linear map

$$f: M \longrightarrow N_{[r]}.$$

Usually we will omit the subscript $[r]$ and simply say that $f: M \to N$ is A-linear.

If $\mathcal{M}$ is an $\mathcal{O}$-module, then the stalk $\mathcal{M}_a$ is naturally an $\mathcal{O}_a$-module. An $\mathcal{O}$-linear map $f: \mathcal{M} \to \mathcal{N}$ between two $\mathcal{O}$-modules is a homomorphism of sheaves of abelian groups such the maps $\mathcal{M}(U) \to \mathcal{N}(U)$ are $\mathcal{O}(U)$-linear. Then the kernel and image also carry natural structures of $\mathcal{O}$-modules. Clearly, the canonical flabby resolution of an $\mathcal{O}$-module is naturally a sequence of $\mathcal{O}$-modules.

Since for every open subset $U \subset X$ we have a ring homomorphism $\mathcal{O}(X) \to \mathcal{O}(U)$, all $\mathcal{M}(U)$ can be considered as $\mathcal{O}(X)$-modules. Hence an $\mathcal{O}$-module can be considered as sheaf of $\mathcal{O}(X)$-modules. In particular, $H^n(X, \mathcal{M})$ carries a natural structure as $\mathcal{O}(X)$-module.

We consider a very special case. We take for $\mathcal{O}$ the sheaf $\mathcal{C}$ of continuous functions. There are two possibilities: $\mathcal{C}_\mathbb{R}$ is the sheaf of continuous real-valued and $\mathcal{C}_\mathbb{C}$ the sheaf of continuous complex-valued functions. If we write $\mathcal{C}$ we mean one of both. The sheaf $\mathcal{C}$ or more generally a module over this sheaf have over paracompact spaces a property which can be considered as a weakend form of flabbyness.

§2. Sheaves of rings and modules

2.1 Remark. *Let X be a paracompact space and let $\mathcal{M}$ be a $\mathcal{C}$-module on X. Assume that U is an open subset and that $V \subset U$ is an open subset whose closure is contained in U. Assume that $s \in \mathcal{M}(U)$ is a section over U. Then there is a global section $S \in \mathcal{M}(X)$ such that $S|V = s|V$.*

Proof. We choose a continuous real valued function φ on X, which is one on V and whose support is contained in U. Then we consider the open covering $X = U \cup U'$, where U' denotes the complement of the support of φ. On U we consider the section φs and on U' the zero section. Since both are zero on $U \cap U'$ they glue to a section S on X. □

2.2 Lemma. *Let X be a paracompact space and let $\mathcal{M} \to \mathcal{N}$ be a surjective $\mathcal{C}$-linear map of $\mathcal{C}$-modules. Then $\mathcal{M}(X) \to \mathcal{N}(X)$ is surjective.*

Proof. Let $s \in \mathcal{N}(X)$. There exists an open covering $(U_i)_{i \in I}$ of X such that $s|U_i$ is the image of a section $t_i \in \mathcal{M}(U_i)$. We can assume that the covering is locally finite. We take open subsets $V_i \subset U_i$ whose closure is contained in U_i and such that (V_i) is still a covering. Then we choose a partition of unity (φ_i) with respect to (V_i). By Remark 2.1 there exists global sections $T_i \in \mathcal{M}(X)$ with the property $T_i|V_i = t_i|V_i$. We now consider

$$T := \sum_{i \in I} \varphi_i T_i.$$

Since I can be infinite, we have to explain what this means. Let $a \in X$ a point. There exists an open neighborhood $U(a)$ such $V_i \cap U(a) \neq \emptyset$ only for a finite subset $J \subset I$. We can define the section

$$T(a) := \sum_{i \in J} \varphi_i T_i | U(a).$$

The sets $U(a)$ cover X and the sections $T(a)$ glue to a section T. Clearly T maps to s. □

2.3 Lemma. *Let X be a paracompact space and let $\mathcal{M} \to \mathcal{N} \to \mathcal{P}$ be an exact sequence of $\mathcal{C}$-modules. Then $\mathcal{M}(X) \to \mathcal{N}(X) \to \mathcal{P}(X)$ is exact too.*

Proof. As mentioned during the proof of Lemma 1.2, it is sufficient to show that a short exact sequence is led to a short exact sequence. The only problem is the surjectivity at the end of the sequence. But this follows from Lemma 2.2. □

Let $\mathcal{M}$ be a $\mathcal{C}$-module over a paracompact space. Then the canonical flabby resolution is also a sequence of $\mathcal{C}$-modules. From 2.3 follows that the resolution remains exact after the application of Γ. We obtain the following proposition.

2.4 Proposition. *Let X be paracompact. Every $\mathcal{C}$-module $\mathcal{M}$ is acyclic, i.e. $H^n(X, \mathcal{M}) = 0$ for $n > 0$.*

3. Čech Cohomology

Here we will consider only the first Čech cohomology group of a sheaf. We have to work with open coverings $\mathfrak{U} = (U_i)_{i \in I}$ of the given topological space X. Let F be a sheaf on X. A *one-cocycle* of F with respect to the covering $\mathfrak{U}$ is a family of sections

$$s_{ij} \in F(U_i \cap U_j), \quad (i,j) \in I \times I,$$

with the following property: for each triple i, j, k of indices one has

$$s_{ik} = s_{ij} + s_{jk} \quad \text{on} \quad U_i \cap U_j \cap U_k.$$

In more precise writing this means

$$s_{ik}|(U_i \cap U_j \cap U_k) = s_{ij}|(U_i \cap U_j \cap U_k) + s_{jk}|(U_i \cap U_j \cap U_k).$$

We denote by $C^1(\mathfrak{U}, F)$ the group of all one-cocycles. Assume that a family of sections $s_i \in F(U_i)$ is given. Then

$$s_{ij} = s_i|(U_i \cap U_j) - s_j|(U_i \cap U_j)$$

obviously is a one-cocycle. We denote it by

$$\delta(s_i)_{i \in I}.$$

A one-cocycle of this form is called a *one-coboundary*. The set of all one-coboundaries is a subgroup

$$B^1(\mathfrak{U}, F) \subset C^1(\mathfrak{U}, F).$$

The (first) Čech cohomology of F with respect to the covering $\mathfrak{U}$ is defined as

$$\check{H}^1(\mathfrak{U}, F) := C^1(\mathfrak{U}, F)/B^1(\mathfrak{U}, F).$$

A homomorphism of sheaves $F \to G$ induces a homomorphism

$$\check{H}^1(\mathfrak{U}, F) \longrightarrow \check{H}^1(\mathfrak{U}, G).$$

Let $f : G \to H$ be a surjective homomorphism of sheaves and $\mathfrak{U} = (U_i)$ an open covering of X. We denote by

$$H_{\mathfrak{U}}(X) = H_{\mathfrak{U},f}(X)$$

the set of all global sections of H with the following property.
For every index i there is a section $t_i \in G(U_i)$ such that $f(t_i) = s|U_i$.

§3. Čech Cohomology

By definition of (sheaf-)surjectivity, for every global section $s \in H(X)$, there exists an open covering $\mathfrak{U}$ with $s \in H_{\mathfrak{U}}(X)$. It follows

$$H(X) = \bigcup_{\mathfrak{U}} H_{\mathfrak{U}}(X).$$

Let $0 \to F \to G \to H \to 0$ be an exact sequence and let $\mathfrak{U}$ be an open covering. There exists a natural homomorphism

$$\delta : H_{\mathfrak{U}}(X) \longrightarrow \check{H}^1(\mathfrak{U}, F),$$

which is constructed as follows. Let be $s \in H_{\mathfrak{U}}(X)$. We choose elements $t_i \in G(U_i)$ which are mapped to $s|U_i$. The differences $t_i - t_j$ come from sections $t_{ij} \in F(U_i \cap U_j)$. They define a one-cocycle $\delta(s)$. It is easy to check that the corresponding element of $\check{H}^1(\mathfrak{U}, F)$ does not depend on the choice of the t_i.

3.1 Lemma. *Let $0 \to F \to G \to H \to 0$ be an exact sequence of sheaves and let $\mathfrak{U}$ be an open covering. The sequence*

$$0 \to F(X) \longrightarrow G(X) \longrightarrow H_{\mathfrak{U}}(X) \overset{\delta}{\longrightarrow} \check{H}^1(\mathfrak{U}, F) \longrightarrow \check{H}^1(\mathfrak{U}, G) \longrightarrow \check{H}^1(\mathfrak{U}, H)$$

is exact.

Remark. *This sequence does not extend naturally to a long sequence.*

The proof is easy, since all maps are given by explicit formulae. □

This Lemma indicates that Čech cohomology must be related to usual cohomology. Another result in this direction gives the following remark.

3.2 Remark. *Let F be a flabby sheaf. Then for every open covering*

$$\check{H}^1(\mathfrak{U}, F) = 0.$$

Proof. We start with a little remark. Assume that the whole space $X = U_{i_0}$ is a member of the covering. Then the Čech cohomology vanishes (for every sheaf): if (s_{ij}) is a one-cocycle, one defines $s_i = s_{i,i_0}$. Then $\delta((s_i)) = (s_{ij})$. For the proof of Remark 3.2 we now consider the sequence

$$0 \longrightarrow F(X) \longrightarrow \prod_i F(U_i) \longrightarrow \prod_{ij} F(U_i \cap U_j) \longrightarrow \prod_{ijk} F(U_i \cap U_j \cap U_k)$$

$$s \longmapsto (s|U_i)$$
$$(s_i) \longmapsto (s_i - s_j)$$
$$(s_{ij}) \longmapsto (s_{ij} + s_{jk} - s_{ik}).$$

We will prove that this sequence is exact. (Then Remark 3.2 follows.) The idea is to sheafify this sequence: For an open subset $U \subset X$ one considers $F|U$

and also the restricted covering $U \cap U_i$. Repeating the above construction for U instead of X one obtains a sequence of sheaves

$$0 \longrightarrow F \longrightarrow \mathcal{A} \longrightarrow \mathcal{B} \longrightarrow \mathcal{C}.$$

Since F is flabby, also $\mathcal{A}, \mathcal{B}, \mathcal{C}$ are flabby. The remark at the beginning of the proof shows that $0 \longrightarrow F(U) \longrightarrow \mathcal{A}(U) \longrightarrow \mathcal{B}(U) \longrightarrow \mathcal{C}(U)$ is exact, when U is contained in some U_i. Hence the sequence is sheaf-exact. From Lemma 1.6 follows that the exactness is also true for $U = X$. □

Let now F be an arbitrary sheaf, $F^{(0)}$ the associated flabby sheaf. We get an exact sequence $0 \to F \to F^{(0)} \to H \to 0$. Let $\mathfrak{U}$ be an open covering. We know that $\check{H}^1(\mathfrak{U}, F^{(0)})$ vanishes (Remark 3.2). From Lemma 3.1 we obtain an isomorphy

$$\check{H}^1(\mathfrak{U}, F) \cong H_\mathfrak{U}(X)/F^{(0)}(X).$$

From the long exact cohomology sequence we get for the usual cohomology

$$H^1(X, F) \cong H(X)/F^{(0)}(X).$$

This gives an *injective* homomorphism

$$\check{H}^1(\mathfrak{U}, F) \longrightarrow H^1(X, F).$$

We obtain the following result.

3.3 Proposition. *Let F be a sheaf. Then*

$$H^1(X, F) = \bigcup_\mathfrak{U} \check{H}^1(\mathfrak{U}, F).$$

The following commutative diagram shows that the homomorphism δ from Lemma 3.1 and that of general sheaf theory Theorem 1.5 coincide:

3.4 Remark. *For a short exact sequence $0 \to F \to G \to H \to 0$ the diagram*

$$\begin{array}{ccccccc}
0 & \longrightarrow & F(X) & \longrightarrow & G(X) & \longrightarrow & H_\mathfrak{U}(X) & \xrightarrow{\delta} & \check{H}^1(\mathfrak{U}, F) \\
& & \| & & \| & & \downarrow & & \downarrow \\
0 & \longrightarrow & F(X) & \longrightarrow & G(X) & \longrightarrow & H(X) & \xrightarrow{\delta} & H^1(X, F)
\end{array}$$

is commutative.

Proof. The Godement resolutions of F, G, H give a short exact sequence of complexes. The groups in the Remark can be expressed explicitly inside this sequence. So the proof can be given by a straight forward computation which can be left to the reader. □

Let $\mathfrak{V} = (V_j)_{j \in J}$ be a refinement of $\mathfrak{U} = (U_i)_{i \in I}$ and $\varphi : J \to I$ a refinement map ($V_\varphi \subset U_i$). Using this refinement map one obtains a natural map

$$\check{H}^1(\mathfrak{U}, F) \longrightarrow \check{H}^1(\mathfrak{V}, F).$$

This shows the following result.

3.5 Remark. *Let $\mathfrak{V}$ be an refinement of $\mathfrak{U}$ and $\varphi : J \to I$ a refinement map. The diagram*

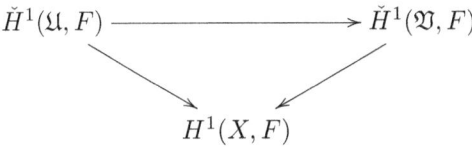

commutes. As a consequence, it doesn't depend on the choice of the refinement map.

Usually it is of course very difficult to control all open coverings of a topological space. (There is also the logical difficulty that the set of all coverings does not exist. It is easy to overcome this difficulty, we omit it.) But sometimes a single covering is sufficient:

3.6 Theorem of Leray. *Let F be a sheaf on X and $\mathfrak{U} = (U_i)$ an open covering of X. Assume that $H^1(U_i, F|U_i) = 0$ for all i. Then*

$$H^1(X, F) = \check{H}^1(\mathfrak{U}, F).$$

Proof. Since two coverings admit a joint refinement, it is sufficient to prove that $\check{H}^1(\mathfrak{U}, F) \to \check{H}^1(\mathfrak{V}, F)$ is an isomorphism for each refinement $\mathfrak{V}$ of $\mathfrak{U}$. Since the map is injective (use Proposition 3.3), it remains to prove surjectivity. We choose a refinement map $\varphi : J \to I$. We denote the indices in I by $i, j, \ldots$ and those of J by $\alpha, \beta, \ldots$. Let be $(s_{\alpha,\beta}) \in C^1(\mathfrak{V}, F)$. We consider the covering $U_i \cap \mathfrak{V} := (U_i \cap V_\alpha)_\alpha$ of U_i. From the assumption $\check{H}^1(U_i \cap \mathfrak{V}, F|U_i) = 0$ we get the existence of $t_{i\alpha} \in F(U_i \cap V_\alpha)$ such that

$$s_{\alpha\beta} = t_{i\alpha} - t_{i\beta} \quad \text{on} \quad U_i \cap V_\alpha \cap V_\beta.$$

From this equation follows that

$$t_{i\alpha} - t_{j\alpha} = t_{i\beta} - t_{j\beta} \quad \text{on} \quad U_i \cap U_j \cap V_\alpha \cap V_\beta.$$

Hence these differences glue to a section $T_{ij} \in F(U_i \cap U_j)$,

$$T_{ij} = t_{i\alpha} - t_{j\alpha} \quad \text{on} \quad U_i \cap U_j \cap V_\alpha.$$

Clearly (T_{ij}) is a one-cocycle in $C^1(\mathfrak{U}, F)$. We consider its image $(T_{\varphi\alpha,\varphi\beta})$ in $C^1(\mathfrak{V}, F)$. It is easy to check that this one-cocycle, and the one we started with $(s_{\alpha\beta})$, defines the same cohomology class: they differ by the one-coboundary $(h_\beta - h_\alpha)$ with $h_\alpha = t_{\varphi\alpha,\alpha} \in F(V_\alpha)$. □

4. Some vanishing results

Let X be a topological space and A an abelian group. We denote by A_X the sheaf of locally constant functions with values in A. This sheaf can be identified with the sheaf that is generated by the presheaf of constant functions. We will write
$$H^n(X, A) := H^n(X, A_X).$$

We mention that these groups under reasonable assumptions (for example for paracompact manifolds) agree with the singular cohomology in the sense of algebraic topology.

4.1 Proposition. *Let U be an open and convex subset of $\mathbb{R}^n$. Then for every abelian group A*
$$H^1(U, A) = 0.$$

This is actually true for all H^n, $n > 0$. The best way to prove this is to use the comparison theorem with singular cohomology as defined in algebraic topology. But we do not want to use this. Therefore we restrict to H^1, where we can give a simple argument.

Proof of 4.1. Every convex open subset of $\mathbb{R}^n$ is topologically equivalent to $\mathbb{R}^n$. Hence it is sufficient to restrict to $U = \mathbb{R}^n$. Just for simplicity we assume $n = 1$. (The general case then should be clear.) We use Čech cohomology. Because of Proposition 3.3 and Remark 3.5 it is sufficient to show that every open covering admits a refinement $\mathfrak{U}$ such that $\check{H}^1(\mathfrak{U}, A_X) = 0$. To show this we take a refinement of a very simple nature. It is easy to show that there exists a refinement of the following form. The index set is $\mathbb{Z}$. There exists a sequence of real numbers (a_n) with the following properties:

a) $a_n \leq a_{n+1}$
b) $a_n \to +\infty$ for $n \to \infty$ and $a_n \to -\infty$ for $n \to -\infty$
c) $U_n = (a_n, a_{n+2})$.

Assume that $s_{n,m}$ is a one-cocycle with respect to this covering. Notice that U_n has non empty intersection only with U_{n-1} and U_{n+1}. Hence only $s_{n-1,n}$ is of relevance. This is a locally constant function on $U_{n-1} \cap U_n = (a_n, a_{n+1})$. Since this is connected, the function $s_{n-1,n}$ is constant. We want to show that it is a one-coboundary, i.e. we want to construct constant functions s_n on U_n such that $s_{n-1,n} = s_n - s_{n-1}$ on (a_n, a_{n+1}). This is easy. One starts with $s_0 = 0$ and then constructs inductively $s_1, s_2, \ldots$ and in the same way for negative n. □

Consider on the real line $\mathbb{R}$ the sheaf of complex valued differentiable functions $\mathcal{C}^\infty$. Taking derivatives one gets a sheaf homomorphism $\mathcal{C}^\infty \to \mathcal{C}^\infty$,

§4. Some vanishing results

$f \mapsto f'$. The kernel is the sheaf of all locally constant functions, which we denote simply by $\mathbb{C}$. Hence we get an exact sequence

$$0 \longrightarrow \mathbb{C} \longrightarrow \mathcal{C}^\infty \longrightarrow \mathcal{C}^\infty \longrightarrow 0.$$

This sequence is exact since every differentiable function has an integral. Hence this sequence can be considered as an acyclic resolution of $\mathbb{C}$. We obtain $H^q(\mathbb{R}, \mathbb{C}) = 0$ for all $q > 0$. For $q = 1$ this follows already from Proposition 4.1. There is a generalization to higher dimensions. For example, a standard result of vector analysis states in the case $n = 2$.

4.2 Lemma. *Let $E \subset \mathbb{R}^2$ be an open and convex subset and let $f, g \in \mathcal{C}^\infty(E)$ be a pair of differentiable functions with the property*

$$\frac{\partial f}{\partial y} = \frac{\partial g}{\partial x}.$$

Then there is a differentiable function h with the property

$$f = \frac{\partial h}{\partial x}, \quad g = \frac{\partial h}{\partial y}.$$

In the language of exact sequences this means that the sequence

$$0 \longrightarrow \mathbb{C} \longrightarrow \mathcal{C}^\infty(E) \longrightarrow \mathcal{C}^\infty(E) \times \mathcal{C}^\infty(E) \longrightarrow \mathcal{C}^\infty(E) \longrightarrow 0$$

$$f \longmapsto \left(\frac{\partial f}{\partial x}, \frac{\partial f}{\partial y}\right)$$

$$(f, g) \longmapsto \frac{\partial f}{\partial y} - \frac{\partial g}{\partial x}$$

is exact. When E is not convex, this sequence needs not to be exact. But since every point in $\mathbb{R}^2$ has an open convex neighborhood, the sequence of sheaves

$$0 \longrightarrow \mathbb{C}_X \longrightarrow \mathcal{C}_X^\infty \longrightarrow \mathcal{C}_X^\infty \times \mathcal{C}_X^\infty \longrightarrow \mathcal{C}_X^\infty \longrightarrow 0$$

is exact for every open subset $X \subset \mathbb{R}^2$. This is an acyclic resolution and we obtain the following proposition:

4.3 Proposition. *For convex open $E \subset \mathbb{R}^2$ we have*

$$H^i(E, \mathbb{C}) = 0 \quad \text{for} \quad i > 0.$$

One can of course consider real valued differentiable functions and the same proof shows $H^i(E, \mathbb{R}) = 0$ for $i > 0$.

As an application we prove the following proposition.

4.4 Proposition. *For convex open $E \subset \mathbb{R}^n$ one has*
$$H^2(E, \mathbb{Z}) = 0.$$

Proof. We consider the homomorphism
$$\mathbb{C} \longrightarrow \mathbb{C}^{\boldsymbol{\cdot}}, \qquad z \longmapsto e^{2\pi i z}.$$
The kernel is $\mathbb{Z}$. This can be considered as an exact sequence of sheaves for example on an open convex $E \subset \mathbb{R}^n$. A small part of the long exact cohomology sequence is
$$H^1(E, \mathbb{C}^{\boldsymbol{\cdot}}) \longrightarrow H^2(E, \mathbb{Z}) \longrightarrow H^2(E, \mathbb{C}).$$
Since the first and the third member of this sequence vanish (Propositions 4.1 and 4.3), we get the proof of Proposition 4.4. □

Next we treat an example of complex analysis.

4.5 Lemma of Dolbeault. *Let $E \subset \mathbb{C}$ be an open disk. For every function $f \in \mathcal{C}^\infty(E)$ there exists $g \in \mathcal{C}^\infty(E)$ with*
$$f = \frac{\partial g}{\partial \bar{z}} := \frac{1}{2}\left(\frac{\partial g}{\partial x} + i\frac{\partial g}{\partial y}\right).$$

A proof can be found in [Fo], Satz 13.2. We give a short sketch here. In a first step one restricts to the case where f has compact support. In this case the function g can be constructed as an integral:
$$g(z) = -\frac{1}{\pi} \int_0^{2\pi} \int_0^1 f(z + re^{i\varphi}) e^{-i\varphi} dr d\varphi.$$
One can show that $\partial g / \partial \bar{z} = f$. But this is not trivial. One has to make use of the Theorem of Stokes V.7.8. We do not give the details.

If f has not compact support, one needs an approximation argument. One writes f as limit of a sequence of functions f_n with compact support such that f and f_n coincide for $|z| < 1 - 1/n$. Then one writes $\partial g_n/\partial \bar{z} = f_n$ using the first part of the proof. We have the freedom to add to g_n a polynomial $P_n(z)$. In this way one can get the convergence of the sequence g_n to a solution g. □

The Lemma of Dolbeault is related to an exact sequence of sheaves. Recall that a (real-) differentiable function is analytic (=holomorphic) if and only if $\partial f/\partial \bar{z} = 0$. We denote by $\mathcal{O}$ the sheaf of holomorphic functions. The Lemma of Dolbeault shows that the sequence of sheaves
$$0 \longrightarrow \mathcal{O}_X \longrightarrow \mathcal{C}_X^\infty \xrightarrow{\bar{\partial}} \mathcal{C}_X^\infty \longrightarrow 0$$
is exact for an open subset $X \subset \mathbb{C}$. This is an acyclic resolution of $\mathcal{O}_X$. Applying the Lemma of Dolbeault once more, we get the following result.

§4. Some vanishing results

4.6 Proposition. *Let $E \subset \mathbb{C}$ be an open disk. Then*
$$H^i(E, \mathcal{O}_E) = 0 \quad \text{for} \quad i > 0.$$

We denote by $\bar{\mathbb{C}}$ the Riemann sphere.

4.7 Theorem. *One has*
$$H^1(\bar{\mathbb{C}}, \mathcal{O}_{\bar{\mathbb{C}}}) = 0.$$

For the proof we consider a covering of $\bar{\mathbb{C}}$ by two disks E_1, E_2 such that the intersection is a circular ring $1 < |z| < 2$. By Leray's theorem 3.6 and the vanishing result Proposition 4.6 it is sufficient to show that the Čech cohomology with respect to this covering vanishes. A Čech one-cocycle simply is given by a holomorphic function on the circular ring. We have to show that it can be written as difference $f_1 - f_2$ where f_i is holomorphic on the disk E_i. This is possible by the theory of the *Laurent decomposition*. □

Chapter V. Basic facts about Riemann surfaces

1. Geometric spaces

In the following we denote by $\mathcal{C}(X)$ the set of complex valued continuous functions on the topological space X.

1.1 Definition. *A geometric structure $\mathcal{O}$ on a topological space is a collection of subrings $\mathcal{O}(U) \subset \mathcal{C}(U)$, where U runs through all open subsets, such that the following conditions are satisfied:*

1) *The constant functions are in $\mathcal{O}(U)$.*
2) *If $V \subset U$ are open sets then*

$$f \in \mathcal{O}(U) \Longrightarrow f|V \in \mathcal{O}(V).$$

3) *Let $(U_i)_{i \in I}$ be a system of open subsets and $f_i \in \mathcal{O}(U_i)$ such that*

$$f_i|U_i \cap U_j = f_j|U_i \cap U_j \quad \text{for all} \quad (i,j),$$

then there exists $f \in \mathcal{O}(U)$ where $U = \bigcup_{i \in I} U_i$ with the property

$$f|U_i = f_i \quad \text{for all} \quad i.$$

We see that $\mathcal{O}$ is a sheaf of rings. We call the functions of $\mathcal{O}(U)$ the *distinguished functions*. Conditions two and three mean that to be distinguished is a local property. Our main example at the moment is $X = \mathbb{C}$, where the distinguished functions are the holomorphic functions.

A *geometric space* is a pair $(X, \mathcal{O})$ consisting of a topological space and a geometric structure.

1.2 Definition. *A morphism $f : (X, \mathcal{O}_X) \to (Y, \mathcal{O}_Y)$ of geometric spaces is a continuous map $f : X \to Y$ with the following additional property. If $V \subset Y$ is open and $g \in \mathcal{O}_Y(V)$, then $g \circ f$ is contained in $\mathcal{O}_X(f^{-1}(V))$.*

Quite trivial facts are:

The composition of two morphisms is a morphism.
The identical map $(X, \mathcal{O}) \to (X, \mathcal{O})$ is a morphism.

A morphism $f : (X, \mathcal{O}_X) \to (Y, \mathcal{O}_Y)$ of geometric spaces is called an *isomorphism* if f is topological and if $f^{-1} : (Y, \mathcal{O}_Y) \to (X, \mathcal{O}_X)$ is also a morphism. This means that the rings $\mathcal{O}_X(U)$ and $\mathcal{O}_Y(f(U))$ are naturally isomorphic.

Let $U \subset X$ be an open subset of a geometric space $(X, \mathcal{O})$. We can define the restricted geometric structure $\mathcal{O}|U$ by

$$\mathcal{O}|U(V) := \mathcal{O}(V) \qquad (V \subset U \text{ open}).$$

It is clear that the natural embedding $i : (U, \mathcal{O}_X|U) \hookrightarrow (X, \mathcal{O}_X)$ is a morphism and moreover that a map $f : Y \to U$ from a geometric space $(Y, \mathcal{O}_Y)$ into U is a morphism if and only if $i \circ f$ is a morphism.

A morphism $f : (X, \mathcal{O}_X) \to (Y, \mathcal{O}_Y)$ is called an *open embedding* if it is the composition of an isomorphism $(X, \mathcal{O}_X) \to (U, \mathcal{O}_Y|U)$, $U \subset Y$ open, and the natural injection.

We used the restriction $\mathcal{O}_X|U$ of a geometric structure to an open subset which is a quite trivial construction. For later purpose we mention that this construction can be given for every presheaf F in the same manner. The presheaf $F|U$ is the presheaf on U defined by $(F|U)(V) := F(V)$ for open subsets $V \subset U$ and the same restriction maps. For every $a \in U$ there is a natural isomorphism $F_a \cong (F|U)_a$. If $\mathcal{O}$ is a sheaf of rings, then $\mathcal{O}|U$ is so and if $\mathcal{M}$ is an $\mathcal{O}$-module then $\mathcal{M}|U$ is an $\mathcal{O}|U$-module. We mention that the restriction is compatible with the notion of a generated sheaf.

1.3 Lemma. *The construction of the generated sheaf is compatible with restriction to open subsets in the follwoing sense. Let F be a presheaf on X and $U \subset X$ open subset. Then there is a natural isomorphism*

$$\hat{F}|U \cong \widehat{F|U}$$

which is the identity on stalks.

The proof can be left to the reader. □

2. The notion of a Riemann surface

We equip $\mathbb{C}$ and more generally an open subset V with the sheaf of holomorphic functions. The geometric space obtained in this way is denoted by $(V, \mathcal{O}_V)$. A Riemann surface is a geometric space which is locally isomorphic to such a space:

2.1 Definition. *A Riemann surface is a geometric space $(X, \mathcal{O}_X)$ such that for every point there exists an open neighborhood U and an open subset $V \subset \mathbb{C}$ such that the geometric spaces $(U, \mathcal{O}_X|U)$ and $(V, \mathcal{O}_V)$ are isomorphic geometric spaces. We always assume that X is a Hausdorff space with countable basis of the topology.*

There exists the empty Riemann surface. We will always assume that X is not empty if a statement makes otherwise no sense.

Of course $(\mathbb{C}, \mathcal{O}_\mathbb{C})$ is a Riemann surface. An open subspace of a Riemann surface (equipped with the induced geometric structure) is a Riemann surface. In particular, every open subset $U \subset \mathbb{C}$ carries a natural structure of a Riemann surface. If U, V are two open subsets of $\mathbb{C}$, then for a map $f : U \to V$ the following two conditions are equivalent.

a) The map f is analytic in the sense of complex analysis.
b) The map f defines a morphism of geometric spaces $f : (U, \mathcal{O}_U) \to (V, \mathcal{O}_V)$.

This allows us to define:

2.2 Definition. *A map $f : (X, \mathcal{O}_X) \to (Y, \mathcal{O}_Y)$ between Riemann surfaces is called holomorphic or analytic if it is a morphism of geometric spaces.*

A biholomorphic map between Riemann surfaces of course is a bijective map which is analytic in both directions.

A (topological) chart on a Riemann surface X is a topological map from an open subset $U \subset X$ onto an open subset $V \subset \mathbb{C}$. The chart is called analytic if it is moreover biholomorphic, i.e. an isomorphism of geometric spaces $(U, \mathcal{O}_X|U) \to (V, \mathcal{O}_V)$.

Let $\varphi : U \to V$ and $\psi : U' \to V'$ be two charts on X. Then we can consider the topological map

$$\gamma := \psi \circ \varphi^{-1} : \varphi(U \cap U') \to \psi(U \cap U').$$

The notation "$\psi \circ \varphi^{-1}$" is not in accordance with the strong rules of set theory, but we allow this and related notations when it is clear from the context what is meant.

The sets $\varphi(U \cap U')$ and $\psi(U \cap U')$ are open subsets of $\mathbb{C}$. If the charts φ, ψ are analytic then the chart change map γ is biholomorphic.

A set of charts $\varphi : U \to V$ is called an *atlas* of X if the domains of definition U cover X. The set of all analytic charts is an atlas.

Riemann surfaces via charts

Assume that a topological space X and a set $\mathcal{A}$ of two dimensional charts (topological maps from open subsets of X to open subsets of $\mathbb{C}$) is given. We assume that $\mathcal{A}$ is an atlas. We also assume that all chart changes $\gamma = \psi \circ \varphi^{-1}$ ($\varphi, \psi \subset \mathcal{A}$) are biholomorphic. Then we call $\mathcal{A}$ an analytic atlas.

2.3 Remark. *Let $\mathcal{A}$ be an analytic atlas on X. Then there exists a unique structure as Riemann surface $(X, \mathcal{O}_X)$ such that all elements of $\mathcal{A}$ are analytic charts with respect to this structure.*

It should be clear how $\mathcal{O}_X$ has to be defined: A function $f : U \to \mathbb{C}$ for open $U \subset X$ belongs to $\mathcal{O}_X(U)$ if and only if for every chart $\varphi : U_\varphi \to V_\varphi$ the function
$$f_\varphi = f \circ \varphi^{-1} : \varphi(U \cap U_\varphi) \longrightarrow \mathbb{C}$$
is analytic in the usual sense.

The atlas $\mathcal{A}$ then is part of the atlas of all analytic charts of $(X, \mathcal{O}_X)$, but it can be smaller. The atlas of all analytic charts is the unique maximal analytic atlas which contains $\mathcal{A}$.

3. The notion of a differentiable manifold

An n-dimensional topological manifold X is a topological space such that each point has an open neighbourhood which is homeomorphic to some open subset of $\mathbb{R}^n$. In the case $n = 2$ we call this a topological surface. Every Riemann surface has an underlying topological surface. Hence Riemann surfaces should be considered as surfaces with an additional structure. We introduced this additional structure as a sheaf $\mathcal{O}_X$. This approach is very flexible and can be used to consider other geometric structures. For example, one can consider for an open subset $U \subset \mathbb{R}^n$ the set $\mathcal{C}^\infty(U)$ of all differentiable functions in the sense of real analysis. One can take them real- or complex valued. A complex valued function is differentiable if and only if real and imaginary part are differentiable. For our purposes it is better to take them complex valued. Hence from now on we will use the notation $\mathcal{C}^\infty(U)$ for the set of *complex valued differentiable functions*. We then can consider the geometric space $(U, \mathcal{C}_U^\infty)$ where $\mathcal{C}_U^\infty(V) := \mathcal{C}^\infty(V)$. Similar to the definition of a Riemann surface, we define an n-dimensional differentiable manifold $(X, \mathcal{C}_X^\infty)$ to be a geometric space that is locally isomorphic to a $(U, \mathcal{C}_U^\infty)$, $U \subset \mathbb{R}^n$ open. A morphism between differentiable manifolds is just called a differentiable map. Isomorphisms between differentiable manifolds are called diffeomorphisms.

Let X be a Riemann surface and U be an open subset. One can define what it means that a function $f : U \to \mathbb{C}$ is differentiable $(\mathcal{C}^\infty)$. For every point $a \in U$ there exists an analytic chart $\varphi : U_\varphi \to V_\varphi$ such that the transported function $f_\varphi(z) = f(\varphi^{-1}(z))$ is differentiable in the usual sense. It is clear that this defines a structure $\mathcal{C}_X^\infty$ as differentiable surface on X. We call this the underlying differentiable surface of X. Holomorphic maps between Riemann surfaces are differentiable maps for the underlying differentiable surfaces.

One reason that it is useful to consider the underlying differentiable surface of a Riemann surface is that in the differentiable world partitions of unity exist. Here a partition of unity (φ_i) is called differentiable if all functions φ_i are differentiable.

3.1 Proposition. *Let X be a paracompact differentiable manifold. For every locally finite open covering there exists a differentiable partition of unity.*

The same proof as in the case of continuous functions works. □

Hence there is the following variant of Proposition IV.2.4.

3.2 Proposition. *Let X be a paracompact differentiable manifold, then every C_X^∞-module $\mathcal{M}$ is acyclic, $H^n(X, \mathcal{M}) = 0$ for $n > 0$.*

4. Meromorphic functions

We recall the topology of the Riemann sphere. A subset $U \subset \bar{\mathbb{C}} = \mathbb{C} \cup \{\infty\}$ is open if $U \cap \mathbb{C}$ is open in $\mathbb{C}$ and if, in the case $\infty \in U$, there exists $C > 0$ with the property
$$z \in \mathbb{C}, \quad |z| > C \Longrightarrow z \in U.$$
Obviously $\bar{\mathbb{C}}$ is a compact space. The subset $\mathbb{C}$ is open and the induced topology is the usual one. The map
$$\bar{\mathbb{C}} \longrightarrow \bar{\mathbb{C}}, \quad z \longmapsto 1/z \quad (1/0 := \infty, \ 1/\infty := 0),$$
is topological.

Let $U \subset \bar{\mathbb{C}}$ be an open set and $f : U \to \mathbb{C}$ a function. We assume that $\infty \in U$. Then $f(1/z)$ is defined in an open neighborhood of 0. We call f analytic at ∞ if $f(1/z)$ is analytic in an open neighborhood of zero. For any open set we define $\mathcal{O}_{\bar{\mathbb{C}}}(U)$ to be the set of all functions with the following properties.

a) The restriction of f to $U \cap \mathbb{C}$ is analytic in the usual sense.
b) When $\infty \in U$ then f is analytic at ∞.

It is easy to see that $(\bar{\mathbb{C}}, \mathcal{O}_{\bar{\mathbb{C}}})$ is a Riemann surface. We describe two analytic charts which cover $\bar{\mathbb{C}}$. They can be used to introduce $\bar{\mathbb{C}}$ as Riemann surface via charts.

1) $\bar{\mathbb{C}} - \{\infty\} = \mathbb{C} \xrightarrow{\mathrm{id}_\mathbb{C}} \mathbb{C},$
2) $\bar{\mathbb{C}} - \{0\} \longrightarrow \mathbb{C}, \quad z \longmapsto 1/z.$

The chart change map is the biholomorphic map
$$\mathbb{C} - \{0\} \longrightarrow \mathbb{C} - \{0\}, \quad z \longmapsto 1/z.$$

We consider now holomorphic maps $f : X \to \bar{\mathbb{C}}$ of an arbitrary Riemann surface into the Riemann sphere.

§4. Meromorphic functions

4.1 Definition. *A meromorphic function f on a Riemann surface X is an analytic map*
$$f : X \longrightarrow \bar{\mathbb{C}},$$
such that $f^{-1}(\infty)$ is a discrete subset of X.

The constant function $f(z) = \infty$ is an analytic map but not meromorphic. Let $S \subset X$ be a discrete subset and let $f : X - S \to \mathbb{C}$ be a holomorphic function. It may happen that f extends to a meromorphic function on X. Then we call the points in S inessential singularities. Since the extension of f to X is unique, we will use the same letter for it.

We denote by $\mathcal{M}(X)$ the set of all meromorphic functions on X. Let $f, g \in \mathcal{M}(X)$ be two meromorphic functions on X and denote by S the union of the points were f or g has the value ∞. Then we can define on $X - S$ the analytic functions $f + g$ and $f \cdot g$. The points of S are inessential singularities. Hence we have defined
$$f + g, \quad fg \in \mathcal{M}(X).$$

It follows that $\mathcal{M}(X)$ is a ring.

If f is a meromorphic function with discrete zero set, then $1/f$ is an analytic function on the complement of this set. Since the singularities are inessential, it extends as a meromorphic function on X.

When $f \in \mathcal{O}_X(X)$ is an analytic function, the composition with the natural inclusion $\mathbb{C} \hookrightarrow \bar{\mathbb{C}}$ is a meromorphic function. We identify f with this meromorphic function. This means that we can consider $\mathcal{O}_X(X)$ as a subring of $\mathcal{M}(X)$.

We want to show that $\mathcal{M}(X)$ is a field, when X is connected. For this we need a a variant of the principle of analytic continuation.

4.2 Lemma. *Let $f, g : X \longrightarrow Y$ be two analytic maps of a connected Riemann surface X into a Riemann surface Y. Assume that there exists a subset $A \subset X$ which is not discrete and such that f and g agree on A. Then $f = g$.*

Corollary. *Let $f : X \to Y$, X connected, be a non-constant analytic map, then $f^{-1}(y)$ is discrete for every $y \in Y$.*

Corollary. *The set $\mathcal{M}(X)$ of all meromorphic functions on a connected Riemann surface is a field.*

Proof. Consider the set of all cumulation points of the set $\{x \in X; \ f(x) = g(x)\}$. It is sufficient to show that this set is open and closed. Since this is a statement of local nature, one can take analytic charts and reduce the statement to the case where X and Y are open subsets of $\mathbb{C}$. Now one can use the standard principle of analytic continuation. □

5. Ramification points

An analytic map $f : X \to Y$ of Riemann surfaces is called *locally biholomorphic* at a point $a \in X$ if f maps some open neighborhood of a biholomorphically to an open neighborhood of $f(a)$. If this is not the case, a is called a *ramification point*. This notion will only be used when f is not constant on the connected component of X which contains a.

5.1 Remark. Let $f : X \to Y$ be a non-constant holomorphic map of connected Riemann surfaces. The set of ramification points is discrete in X.

Proof. Taking charts, one can assume that X and Y are open subsets of $\mathbb{C}$. The ramification points then are the zeros of the derivative of f. □

We recall a result of complex calculus. Let $f : U \to \mathbb{C}$ be a non-constant analytic function on an open connected neighborhood of 0 with the property $f(0) = 0$. There exists a small open neighborhood $0 \in V \subset U$ and an analytic function $h : V \to \mathbb{C}$ with the properties

$$f(z) = h(z)^n, \qquad h'(0) \neq 0.$$

If V is taken small enough, then h maps V biholomorphically onto an open neighborhood of 0. (The number n is the zero order of f.) We refer to [FB] for a proof. There it has been used for the proof of the Open Mapping Theorem III.3.3.

We want to reformulate this result for Riemann surfaces. For sake of convenience we will use the following notation:

A **disk** around $a \in X$ on a Riemann surface X is a biholomorphic map (=analytic chart)

$$\varphi : U \xrightarrow{\sim} \mathbb{E}, \qquad \varphi(a) = 0.$$

Here

$$\mathbb{E} := \{ q \in \mathbb{C};\ |q| < 1 \}$$

denotes the unit disk.

If b is a point of U, we also say that the disk contains b. If U is a subset of a subset $A \subset X$, then we say that the disk is contained in A. (This is not quite correct since a disk is a map and not only the set U.)

If a is a point of a Riemann surface, then of course there exists a disk around a. One simply takes an arbitrary analytic chart $\varphi : U \to V$, $a \in U$. Then one replaces V by a small disk around $f(a)$ and U by the inverse image. For trivial reasons there exists a biholomorphic map of an arbitrary disk to the unit disk, such that the center goes to the center. One can say a little more:

Around an arbitrary point a of a Riemann surface there exist arbitrary small disks.

This means of course that for a given neighborhood one can find a disk which is contained in it.

§5. Ramification points

5.2 Remark. *Let $f : X \to Y$ be a non-constant analytic map of connected Riemann surfaces and $a \in X$ a given point. There exist disks $\psi : V \to E$ around $f(a) \in Y$ and $\varphi : U \to E$ around a such that $f(U) \subset V$ and such that the diagram*

$$\begin{array}{ccc} U & \xrightarrow{\varphi} & E & & E \\ {\scriptstyle f}\downarrow & & \downarrow & & \downarrow {\scriptstyle q} \\ V & \xrightarrow{\psi} & E & & E \\ & & & & {\scriptstyle q^n} \end{array}$$

commutes for some natural number n which is uniquely determined.

This is just a reformulation of the discussion above. □

A simple way to express Remark 5.2 is:

Analytic maps of Riemann surfaces look locally like "$q \mapsto q^n$".

The point a is a ramification point if and only if $n > 1$.

Proper maps

For *proper* analytic maps $f : X \to Y$ there are better results. We will use in the following frequently the following trivial fact: Let $V \subset Y$ an open subset and $U = f^{-1}(V)$ its (full!) inverse image. Then $f : U \to V$ is proper too (Lemma I.3.1).

The basic fact which we have to use is the following purely topological result.

5.3 Proposition. *Let $f : X \to \mathbb{E}^{\bullet}$ be a locally topological proper map of a connected Hausdorff space into the punctured unit disk $\mathbb{E}^{\bullet} = \mathbb{E} - \{0\}$. Then there exists a natural number n and a topological map $\sigma : X \to \mathbb{E}^{\bullet}$ such that the diagram*

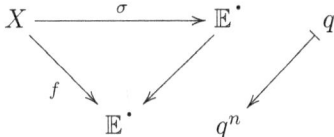

commutes. The number n is uniquely determined.

The proof uses topological covering theory. A quick proof can be found in [Fr1], Chapt. I, Sect 3, Appendix A. We will not give the details of the proof here. We just mention the topological background. Locally, topological and proper maps are special cases of (unramified) topological coverings. These are classified by the fundamental group of the space. We will introduce the fundamental group in Chap. VII, Sect. 3. The fundamental group of $\mathbb{E}^{\bullet}$ is isomorphic to $\mathbb{Z}$. This implies that for each natural number n there exists only one covering of $\mathbb{E}^{\bullet}$ of degree n. □

We point out as a consequence the following result.

5.4 Corollary of 5.3. *Let, in addition, X be a Riemann surface and f analytic. Then the map σ is biholomorphic.*

A simple consequence of Proposition 5.3 states:

5.5 Lemma. *Let $f : X \to \mathbb{E}$, $a \mapsto 0$, be a holomorphic and proper map of a connected Riemann surface X into the unit disk which is locally biholomorphic outside a. Then there exists a biholomorphic map $\sigma : X \to \mathbb{E}$ and a natural number n such that the diagram*

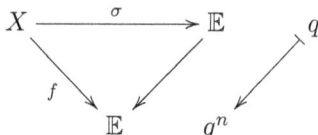

commutes.

Proof. From Proposition 5.3 follows the existence of a biholomorphic map $\sigma : X - \{a\} \to \mathbb{E}^{\bullet}$. We extend it by $\sigma(a) = 0$. The Riemann extension theorem shows that this map is still holomorphic. The map $X \to \mathbb{E}$ is bijective and holomorphic. This implies that it is biholomorphic (compare [FB], IV.4.2). □

We treat a topological application of Proposition 5.3. We take (in the notation of Proposition 5.3) a symbol a which is not contained in X. We define $\bar{X} = X \cup \{a\}$. We extend the map σ to the map

$$\bar{\sigma} : \bar{X} \longrightarrow \mathbb{E}, \quad \bar{\sigma}(a) = 0.$$

This is a bijective map. We equip $\bar{X}$ with the unique topology such that this map is topological. Then X is an open subspace of $\bar{X}$. We also extend f to a map

$$\bar{f} : \bar{X} \longrightarrow \mathbb{E}, \quad \bar{f}(a) = 0.$$

This map still is continuous, even proper but usually not locally topological.

This trivial construction admits an important extension:

5.6 Proposition. *Let $\bar{Y}$ be a surface and let $Y \subset \bar{Y}$ be an open subset such that $S := \bar{Y} - Y$ is a finite set. Assume that $f : X \to Y$ is a locally topological proper map of a Hausdorff space X into Y. Then there exists a surface $\bar{X}$ with the following property:*

a) *X is an open subset of $\bar{X}$ and the topology of X is the induced topology.*
b) *$T := \bar{X} - X$ is a finite set. The map f extends to a continuous and proper map*

$$\bar{f} : \bar{X} \longrightarrow \bar{Y}.$$

§5. Ramification points

c) *For every point $a \in T$ there exist neighborhoods $a \in U \subset \bar{X}$ and $\bar{f}(a) \in V \subset \bar{Y}$ and topological maps $U \to \mathbb{E}$ and $V \to \mathbb{E}$ and a natural number n such that the diagram*

$$\begin{array}{ccc} U & \longrightarrow & \mathbb{E} \\ \bar{f}\downarrow & & \downarrow \\ V & \longrightarrow & \mathbb{E} \end{array} \qquad \begin{array}{c} q \\ \downarrow \\ q^n \end{array}$$

commutes.

Additional Remark. *When $\bar{Y}$ is compact, then so is $\bar{X}$.*

Proof. For each $b \in S$ we choose an open connected neighborhood $V(b)$ and a topological map $V(b) \to \mathbb{E}$, $b \mapsto 0$. We can assume that the $V(b)$ for different b are disjoint. Next we take the inverse image $f^{-1}(V(b))$ in X. This needs not to be connected. Hence we consider the connected components. From the properness of f follows that there are only finitely many connected components. Let U be one of them. The map

$$U \longrightarrow V(b) - \{b\} \cong \mathbb{E}^{\bullet}$$

still is proper. To this situation we apply the above construction adding one extra point to U. To be precise we take for each point $b \in S$ and for each connected component of $f^{-1}(V(b))$ an extra symbol $a_{U,b}$. Then we add to U this symbol to get $\bar{U} = U \cup \{a_{U,b}\}$. Now we consider

$$\bar{X} = X \cup \bigcup_{b \in S,\, U} \{a_{U,b}\}.$$

Then $\bar{U}$ is a subset of $\bar{X}$. It is clear how to extend f to a map $\bar{f} : \bar{X} \to \bar{Y}$. It is also clear how to define the topology on $\bar{X}$: a subset is called open if the intersection with X and all $\bar{U}$ is open. □

If $f : U \to \mathbb{C}$ is a non-constant function on a connected open subset $U \subset \mathbb{C}$, then, for $a \in U$, the zero order of $f(z) - f(a)$ at a is called the multiplicity of f at a. If $f : X \to Y$ is a non-constant holomorphic map of Riemann surfaces, then the multiplicity of f at a point $a \in X$ can defined in an obvious way via charts. It also can be defined by means of Remark 5.2.

5.7 Definition. *Let $f : X \to Y$ be a non-constant holomorphic map of connected Riemann surfaces. The multiplicity $\mathrm{Ord}(f, a)$ of f at a point $a \in X$ is the number n such that f looks locally at a like $q \mapsto q^n$ (see Remark 5.2).*

From Lemma 5.5 and the above discussion we deduce the following proposition.

5.8 Proposition. *Let $f : X \to Y$ be a proper non-constant holomorphic map of connected Riemann surfaces. Denote for $b \in Y$ by d the number of all $a \in X$ with $f(a) = b$ counted with multiplicity. The number d is independent of b. In particular, f is surjective.*

We call d the *covering degree* of f.

Proof of Proposition 5.8. The statement is trivial for the map $\mathbb{E} \to \mathbb{E}$, $q \mapsto q^n$. The covering degree in this case is n. We claim that the function that associates to $b \in Y$ the number d is locally constant. It is sufficient to prove this in the case $Y = \mathbb{E}$. In this case we can apply Lemma 5.5 to the connected components of X. So we have proved that the function $b \mapsto d$ is locally constant. Since Y is connected, it is constant. □

6. Examples of Riemann surfaces

The only Riemann surfaces which we introduced so far is the Riemann sphere and its open subsets. We give some more interesting examples but keep short, since they are not needed for the development of the general theory. We will treat now three examples, complex tori, the Riemann surface of an analytic function, and the compact Riemann surfaces that can be associated to algebraic functions. In the last Chapter VIII we will introduce another important example, namely the Riemann surface that can be associated to a discrete subgroup of $\mathrm{SL}(2, \mathbb{R})$.

Tori

Let $L \subset \mathbb{C}$ be a lattice, i.e. $L = \mathbb{Z}\omega_1 + \mathbb{Z}\omega_2$, where ω_1, ω_2 is an $\mathbb{R}$-basis of $\mathbb{C}$. The quotient $X_L = \mathbb{C}/L$ carries a natural structure as Riemann surface. The topology is the quotient topology. A function $f : U \to \mathbb{C}$ on an open subset $U \subset X_L$ is called holomorphic if and only if the composition with the natural projection $p : \mathbb{C} \to X_L$ is a holomorphic function in the usual sense on $p^{-1}(U)$. The natural projection p is holomorphic and even more locally biholomorphic. It follows that the meromorphic functions $f : X_L \to \bar{\mathbb{C}}$ are in 1-1-correspondence to the elliptic functions $F = f \circ p$ (meromorphic functions on $\mathbb{C}$ which are periodic with respect to L).

It can be shown that two tori X_{L_1} and X_{L_2} are biholomorphic equivalent if and only if there exists a complex number $a \in \mathbb{C}$ such that $L_2 = aL_1$. Since this is usually not the case and since two tori are always topologically equivalent, we see that topologically equivalent Riemann surfaces are usually not biholomorphic equivalent.

§6. Examples of Riemann surfaces

Concrete Riemann surfaces

A function element (a, P) is a point $a \in \mathbb{C}$ together with an element $P \in \mathcal{O}_{\mathbb{C},a}$. We can think of P as a power series with center a and positive convergence radius. Let $\alpha : [0, 1] \to \mathbb{C}$ be a curve with starting point $a = \alpha(0)$. Assume that a function element (a, P) is given. We say that (a, P) admits analytic continuation along α if there exists for every $t \in [0, 1]$ a function element $(\alpha(t), P_t)$ such that the following conditions hold:

a) $(a, P) = (\alpha(0), P_0)$.
b) Let $t \in [0, 1]$. If $t' \in [0, 1]$ is close enough to t, then the (open) convergence disks of P_t and $P_{t'}$ have non empty intersection and both functions agree in the intersection.

It is easy to see that such an analytic continuation is unique. (Consider for two continuations P_t, Q_t the supremum of all t such that the function elements agree on $[0, t]$.) Two function elements (a, P) and (b, Q) are called equivalent if there exists a path α with $\alpha(0) = a$ and $\alpha(1) = b$ such that (a, P) can be continued analytically along α with the end-element (b, Q).

The basic truth is that (b, Q) depends on the choice of the curve α. As an example one considers the function element $(1, \sqrt{z})$, where $\sqrt{z}$ is defined by the principal part of the logarithm. This function element can be continued to -1. But if one takes the continuation along the half circle in the upper half plane, one gets a different result as if one takes the half circle in the lower half plane. The two continuations differ by a sign.

The idea of the Riemann surface is to count the point -1 twice. This means that one takes a two-fold covering of $\mathbb{C}^{\bullet}$. On this covering $\sqrt{z}$ can be defined as an unambiguous function.

To define this in a mathematical rigorous way, one starts with an analytic function $f : D \subset \mathbb{C}$ on some connected open subset of $\mathbb{C}$. The elements (a, f_a) of course all are equivalent. Now we introduce the set $\mathcal{R} = \mathcal{R}(f)$ of all function elements which are equivalent to the elements (a, f_a). So $\mathcal{R}$ contains all possible analytic continuations of f.

We introduce a topology on $\mathcal{R}$. Let $(a, P) \in \mathcal{R}$. Consider a positive number ε which is smaller than the convergence radius of P. Then we define

$$U_\varepsilon(a, P) := \{ (b, Q); \ b \in U_\varepsilon(a), \ Q \text{ is the germ of } P \text{ in } b \}.$$

A subset $U \subset \mathcal{R}$ is called open if for every point $a \in U$ there exists a small $\varepsilon > 0$ such that

$$U_\varepsilon(a, P) \subset U.$$

It is clear that this is a (Hausdorff) topology on $\mathcal{R}$. The natural projection

$$p : \mathcal{R} \longrightarrow \mathbb{C}, \quad (a, P) \longmapsto a,$$

is continuous and moreover the restriction

$$U_\varepsilon(a, P) \longrightarrow U_\varepsilon(a) \qquad (\varepsilon \text{ small enough})$$

is topological. Hence we see that the map $\mathcal{R} \to \mathbb{C}$ is locally topological. This is enough information to equip $\mathcal{R}$ with a structure as Riemann surface.

6.1 Remark. *Let Y be a Riemann surface and $f : X \to Y$ a locally topological map of a Hausdorff space X into Y. Then X carries a unique structure as Riemann surface such that f is holomorphic.*

Proof. One defines $\mathcal{O}_X(U)$ to be the set of all functions $f : U \to \mathbb{C}$ such that for all $U_\varepsilon(a, P) \subset U$ the composition

$$U_\varepsilon(a) \xrightarrow{\sim} U_\varepsilon(a, P) \xrightarrow{f} \mathbb{C}$$

is analytic in the usual sense. It is easy to verify the demanded properties. □

Besides the projection $\mathcal{R} \to \mathbb{C}$, $(a, P) \mapsto a$, one can consider the map

$$F : \mathcal{R} \longrightarrow \mathbb{C}, \quad F(a, P) = P(a).$$

This function is of course analytic. There is a natural map

$$D \longrightarrow \mathcal{R}, \quad a \longmapsto (a, f_a),$$

which obviously is an open imbedding. The basic fact is that the commutative diagram

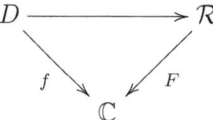

shows F to be an extension of the original function f. This extension includes all possible analytic continuations of f. In a naive sense they are a multi-valued function. The idea of the Riemann surface $\mathcal{R}$ is to obtain a single-valued function on a covering of $\mathbb{C}$.

The Riemann surface of an algebraic function

We show how a compact Riemann surface can be associated to an algebraic function. We will keep rather short. More details can be found in [Fr1], Chapt. I, Sect. 3.

Let $P(z, w)$ be a polynomial of two variables. We assume in the following that P is irreducible. This means that P cannot be written as product of two non-constant polynomials. We also assume that P depends on both variables properly.

The "algebraic function" f related to P is the solution of $P(z, f(z)) = 0$. This is not a function in the usual sense, since it is multi-valued. The origin of the theory of Riemann surfaces can be seen in the wish to get a precise definition of f as a single valued function on a suitable surface. One could use the technique of concrete Riemann surfaces to construct such a surface. We prefer to use the algebraic curve related to P. In both approaches the main problem is to get a *compact* connected

§6. Examples of Riemann surfaces

Riemann surface. One can show that every compact and connected Riemann surface is biholomorphic equivalent to one which comes from an algebraic function. Hence compact Riemann surfaces and algebraic functions describe the same objects.

The affine curve associated to P is defined as

$$\mathcal{N} := \{\, (z, w) \in \mathbb{C} \times \mathbb{C}; \quad P(z, w) = 0 \,\}.$$

6.2 Lemma. *The fibres of the map $p : \mathcal{N} \to \mathbb{C}$, $p(z, w) = z$, are finite.*

This is an easy application of the fact that a non-zero polynomial in one variable has only finitely many zeros. □

6.3 Proposition. *There are finite subsets $S \subset \mathbb{C}$ and $T = p^{-1}(S)$ such that*

$$p : \mathcal{N} - T \longrightarrow \mathbb{C} - S$$

is proper and locally topological.

We just describe the set S. A point $a \in \mathbb{C}$ is contained in S if one of the following conditions is satisfied:

a) There is a b such that P and $\partial P / \partial w$ both vanish at (a, b).
b) Write P in the form

$$P(z, w) = a_0(z) + \cdots + a_n(z) w^n$$

such that the polynomial a_n does not vanish identically. Then $a_n(a) = 0$.

We do not give more details and mention just that b) implies that $p : \mathcal{N} - T \longrightarrow \mathbb{C} - S$ is locally topological. (One has to use the theorem of implicit functions.) That p is proper uses a). □

Now we imbed $\mathbb{C} - S$ into the Riemann sphere $\bar{\mathbb{C}}$ and apply Proposition 5.6.

6.4 Theorem. *There exists a **compact** Riemann surface X which contains $\mathcal{N} - T$ as open sub-surface (with the induced topology and the restricted geometric structure). The complement is a finite set. The map $p : \mathcal{N} - T \longrightarrow \mathbb{C} - S$ extends to a meromorphic function*

$$p : X \longrightarrow \bar{\mathbb{C}}.$$

We also mention a result, which often is clear in concrete situations but needs a proof in general. We refer to [Fr1], Proposition I.3.10, for a proof.

6.5 Theorem. *The compact Riemann surface X attached to an algebraic function is connected. The second projection $q(z,w) = w$ extends also to a meromorphic function*
$$q : X \longrightarrow \bar{\mathbb{C}}.$$

We end this section with a special example in which the above results can be proved more or less easily. Let $Q(z)$ be a non-constant polynomial in one variable without multiple zeros and take $P(z,w) = w^2 - Q(z)$. It is easy to show that P and $\partial P/\partial w$ have no common zeros. It follows that $\mathcal{N}$ is already a surface (an imbedded manifold in the sense of analysis). It carries a structure as Riemann surface such that $p : \mathcal{N} \to \mathbb{C}$ is proper and holomorphic. The covering degree is two. The compactification $\mathcal{N} \subset X$ needs one or two additional points. If it is one, the map p is locally of the form $z \mapsto z^2$. If there are two, the map is locally biholomorphic at both. Later we will see that the compactification by one (resp. two) points depends on whether the degree of Q is odd (resp. even).

Recall that we have two projections $p, q : X \to \bar{\mathbb{C}}$ which are induced by $p(z,w) = z$ and $q(z,w) = w$. The first projection plays the role of the natural "coordinate" on X. Hence we write simply z instead of p. The second projection q describes the solution of the equation $w^2 = Q(z)$. Hence we write simply $\sqrt{Q(z)}$ for q. We see that the a priori "double valued function" $\sqrt{Q(z)}$ appears as usual single-valued function q on a two fold covering of $\bar{\mathbb{C}}$.

7. Differential forms

Let $U \subset \mathbb{C}$ be an open subset. We consider the space $\mathcal{A}^p(U)$ ($p = 0, 1, 2$) of complex valued C^∞-differential forms,
$$\mathcal{A}^0(U) = C^\infty(U), \quad \mathcal{A}^1(U) = C^\infty(U) \times C^\infty(U), \quad \mathcal{A}^2(U) = C^\infty(U).$$
As usually we write the elements of $\mathcal{A}^1(U)$ in the form $f dx + g dy$ and the elements of $\mathcal{A}^2(U)$ in the form $f dx \wedge dy$. We recall the operators
$$\mathcal{A}^1(U) \times \mathcal{A}^1(U) \longrightarrow \mathcal{A}^2(U),$$
$$(f_1 dx + g_1 dy) \wedge (f_2 dx + g_2 dy) = (f_1 g_2 - f_2 g_1) dx \wedge dy,$$
(exterior product), regulated by the the rules $dx \wedge dx = dy \wedge dy = 0$, $dx \wedge dy = -dy \wedge dx$ and the exterior derivatives
$$d : \mathcal{A}^0(U) \longrightarrow \mathcal{A}^1(U), \quad df = \frac{\partial f}{\partial x} + \frac{\partial f}{\partial y}$$
and
$$d : \mathcal{A}^1(U) \longrightarrow \mathcal{A}^2(U), \quad d(f dx + g dy) = \left(\frac{\partial g}{\partial x} - \frac{\partial f}{\partial y}\right) dx \wedge dy.$$
One has $d \circ d = 0$. From Lemma IV.4.2 we obtain the following result.

§7. Differential forms

7.1 Lemma of Poincaré.
The sequence
$$0 \longrightarrow \mathbb{C} \longrightarrow \mathcal{A}^0(U) \longrightarrow \mathcal{A}^1(U) \longrightarrow \mathcal{A}^2(U) \longrightarrow 0$$
is exact for convex U.

We use the notations
$$dz = dx + \mathrm{i}dy, \qquad d\bar{z} = dx - \mathrm{i}dy$$
and
$$\frac{\partial}{\partial z} := \frac{1}{2}\left(\frac{\partial}{\partial x} - \mathrm{i}\frac{\partial}{\partial y}\right), \quad \frac{\partial}{\partial \bar{z}} := \frac{1}{2}\left(\frac{\partial}{\partial x} + \mathrm{i}\frac{\partial}{\partial y}\right).$$

Notice also
$$dz \wedge d\bar{z} = -d\bar{z} \wedge dz = -2\mathrm{i}dx \wedge dy.$$
We can write a one-form also in the form $fdz + gd\bar{z}$.

7.2 Lemma.
The operator d can be rewritten as
$$df = \frac{\partial f}{\partial z}dz + \frac{\partial f}{\partial \bar{z}}d\bar{z} \quad \text{and} \quad d(fdz + gd\bar{z}) = \frac{\partial f}{\partial \bar{z}}d\bar{z} \wedge dz + \frac{\partial g}{\partial z}dz \wedge d\bar{z}.$$

The easy proof is omitted. □

We obtain a splitting
$$\mathcal{A}^1(U) = \mathcal{A}^{1,0}(U) \oplus \mathcal{A}^{0,1}(U)$$
with
$$\mathcal{A}^{1,0}(U) = \mathcal{C}^\infty(U)dz \quad \text{and} \quad \mathcal{A}^{0,1}(U) = \mathcal{C}^\infty(U)d\bar{z}.$$
A one-differential is called holomorphic if it is of the form fdz with holomorphic f. From the Cauchy–Riemann equations follows
$$df = f'(z)dz \quad \text{for holomorphic } f.$$
Hence df is holomorphic for holomorphic f. The set of all holomorphic one-forms is denoted by $\Omega(U)$. It is a subspace of $\mathcal{A}^{1,0}(U)$. We also introduce the operators
$$\partial : \mathcal{A}^0(U) \longrightarrow \mathcal{A}^{1,0}(U), \qquad \partial(f) := \frac{\partial f}{\partial z}dz,$$
$$\bar{\partial} : \mathcal{A}^0(U) \longrightarrow \mathcal{A}^{0,1}(U), \qquad \bar{\partial}(f) := \frac{\partial f}{\partial \bar{z}}d\bar{z}$$
So we have
$$df = \partial f + \bar{\partial} f.$$
Similarly one can define
$$\partial : \mathcal{A}^{0,1}(U) \longrightarrow \mathcal{A}^2(U), \qquad \partial(fd\bar{z}) := \frac{\partial f}{\partial z}dz \wedge d\bar{z},$$
$$\bar{\partial} : \mathcal{A}^{1,0}(U) \longrightarrow \mathcal{A}^2(U), \qquad \bar{\partial}(fdz) := \frac{\partial f}{\partial \bar{z}}d\bar{z} \wedge dz.$$
But these are not really new operators since Lemma 7.2 shows
$$\partial(fd\bar{z}) = d(fd\bar{z}), \quad \bar{\partial}(fdz) = d(fdz).$$
We can reformulate the Lemma of Dolbeault IV.4.5 as follows.

7.3 Lemma of Dolbeault. *For a disk U the sequence*

$$0 \longrightarrow \mathcal{O}(U) \longrightarrow \mathcal{A}^0(U) \xrightarrow{\bar{\partial}} \mathcal{A}^{0,1}(U) \longrightarrow 0$$

is exact, or equivalently,

$$0 \longrightarrow \Omega(U) \longrightarrow \mathcal{A}^{1,0}(U) \xrightarrow{\bar{\partial}} \mathcal{A}^2(U) \longrightarrow 0$$

is exact.

Transformation of differential forms

Let $\gamma = \gamma_1 + i\gamma_2 : U \to V$ be a C^∞-map of open subsets $U, V \subset \mathbb{C}$. One defines the pull-back

$$\gamma^* : \mathcal{A}^p(V) \longrightarrow \mathcal{A}^p(U)$$

as follows:

a) for 0-forms: $\gamma^*(f) = f \circ \gamma$,
b) for 1 forms: $\gamma^*(fdx + gdy) = (f \circ \gamma)\, d\gamma_1 + (f \circ \gamma)\, d\gamma_2$,
c) for 2-forms: $\gamma^*(fdx \wedge dy) = (f \circ \gamma) \left(\dfrac{\partial \gamma_1}{\partial x} \dfrac{\partial \gamma_2}{\partial y} - \dfrac{\partial \gamma_1}{\partial y} \dfrac{\partial \gamma_2}{\partial x} \right) dx \wedge dy$.

The chain rule says: $\gamma_1^* \circ \gamma_2^* = (\gamma_2 \circ \gamma_1)^*$. Another consequence of the chain rule is that the derivative d is compatible with transformation:

$$\gamma^*(d\omega) = d(\gamma^*\omega).$$

We also mention

$$\gamma^*(\omega \wedge \omega') = (\gamma^*\omega) \wedge (\gamma^*\omega')$$

which is easy to verify.

Now we assume that γ is holomorphic. We denote the variable in U by z and in V by w. Then we have

$$\gamma^*(dw) = d(\gamma) = \partial\gamma = \frac{\partial \gamma}{\partial z} dz.$$

Hence for holomorphic γ the transformation gets the simple form

$$\boxed{\gamma^*(f(w)dw) = f(\gamma z)\gamma'(z)dz.}$$

In the same way one proves

$$\gamma^*(d\bar{w}) = \frac{\partial \bar{\gamma}}{\partial \bar{z}} d\bar{z}.$$

This shows the following result.

§7. Differential forms

7.4 Lemma. Let $\gamma : U \to V$ be a **holomorphic** map between open subsets of $\mathbb{C}$. Then γ^* preserves the types $\mathcal{A}^{0,1}$ and $\mathcal{A}^{1,0}$ and it maps holomorphic forms to holomorphic forms. So γ^* gives maps

$$\mathcal{A}^{1,0}(V) \longrightarrow \mathcal{A}^{1,0}(U), \quad \mathcal{A}^{0,1}(V) \longrightarrow \mathcal{A}^{0,1}(U), \quad \Omega(V) \longrightarrow \Omega(U).$$

Differential forms on Riemann surfaces

By a p-form on a Riemann surface one understands a family of p-forms $\omega_\varphi \in \mathcal{A}^p(V_\varphi)$, where $\varphi : U_\varphi \to V_\varphi$ runs through all analytic charts, such that the following condition is satisfied.

If φ, ψ are two analytic charts and $\gamma = \psi \circ \varphi^{-1}$ is the coordinate change map, then

$$\omega_\psi = \gamma^* \omega_\varphi$$

holds on the open subsets where γ is defined. If one has a family, which is only defined for all φ from an atlas of analytic charts, then this family extends uniquely to the atlas of all analytic charts. All what one has to use is the chain rule $\gamma_1^* \circ \gamma_2^* = (\gamma_2 \circ \gamma_1)^*$ and the fact that transformation and derivatives are compatible with restriction to smaller open subsets.

In particular, a zero form is a family of functions $f_\varphi : V_\varphi \to \mathbb{C}$ such that the transported functions in U_φ coincide in the intersections. Hence they glue to a function $f : X \to \mathbb{C}$. We will identify f with the family (f_φ). We denote by $\mathcal{A}^p(X)$ the space of p-forms. We have $\mathcal{A}^0(X) = C^\infty(X)$.

A 1-form is called holomorphic if all its components ω_φ are holomorphic. It is enough to demand this for an atlas of analytic charts, since the transform of a holomorphic 1-form by a holomorphic transformation is holomorphic again. (Here we make use of the fact that we have a Riemann surface and not only a differentiable surface in the sense of real analysis). We denote by $\Omega(X)$ the space of holomorphic 1-forms.

Another obvious property of *holomorphic* transformations is that it preserves the types $\mathcal{A}^{0,1}$ and $\mathcal{A}^{1,0}$. This also follows from the Cauchy Riemann differential equations. Hence we can define $\mathcal{A}^{0,1}(X)$ and $\mathcal{A}^{1,0}(X)$ for a Riemann surface componentwise. The operators d and ∂ commute with holomorphic transformations. Hence we can define operators

$$d : C^\infty(X) \longrightarrow \mathcal{A}^1(X), \quad d : \mathcal{A}^1(X) \longrightarrow \mathcal{A}^2(X),$$
$$\bar{\partial} : C^\infty(X) \longrightarrow \mathcal{A}^{0,1}(X), \quad \bar{\partial} : \mathcal{A}^{1,0}(X) \longrightarrow \mathcal{A}^2(X)$$

also componentwise (for example $(d\omega)_\varphi := d(\omega_\varphi)$.)

Also the exterior product generalizes to Riemann surfaces via charts.

Sheaves of differential forms

If one attaches to each open subset U of a Riemann surface the various spaces of differential forms, one obtains sheaves which we denote by $\mathcal{A}_X^p$, $\mathcal{A}_X^{0,1}$, $\mathcal{A}_X^{1,0}$, Ω_X. We also obtain sequences of sheaves:

$$0 \longrightarrow \mathbb{C}_X \longrightarrow \mathcal{C}_X^\infty \longrightarrow \mathcal{A}_X^1 \longrightarrow \mathcal{A}_X^2 \longrightarrow 0,$$

$$0 \longrightarrow \mathcal{O}_X \longrightarrow \mathcal{C}_X^\infty \overset{\bar{\partial}}{\longrightarrow} \mathcal{A}_X^{0,1} \longrightarrow 0,$$

$$0 \longrightarrow \Omega_X \overset{\text{inclusion}}{\longrightarrow} \mathcal{A}_X^{1,0} \overset{\bar{\partial}}{\longrightarrow} \mathcal{A}_X^2 \longrightarrow 0.$$

The Lemma of Poincaré 7.1 and the Lemma of Dolbeault 7.3 imply that these sequences of sheaves are exact. Hence we obtain the following two theorems:

7.5 Theorem of de Rham. *Let X be a connected Riemann surface. Then*

$$H^0(X, \mathbb{C}) = \mathbb{C}, \quad H^1(X, \mathbb{C}) = \frac{\text{Kernel}(\mathcal{A}^1(X) \longrightarrow \mathcal{A}^2(X))}{\text{Image}(\mathcal{C}^\infty(X) \longrightarrow \mathcal{A}^1(X))}$$

$$H^2(X, \mathbb{C}) = \frac{\mathcal{A}^2(X)}{\text{Image}(\mathcal{A}^1(X) \longrightarrow \mathcal{A}^2(X))}, \quad H^q(X, \mathbb{C}) = 0 \ \text{for } q > 2.$$

7.6 Theorem of Dolbeault. *Let X be a connected Riemann surface. Then*

$$H^1(X, \mathcal{O}_X) = \frac{\mathcal{A}^{0,1}(X)}{\text{Image}(\mathcal{C}^\infty(X) \longrightarrow \mathcal{A}^{0,1}(X))}, \quad H^q(X, \mathcal{O}_X) = 0 \ \text{for } q > 1,$$

and

$$H^1(X, \Omega_X) = \frac{\mathcal{A}^2(X)}{\text{Image}(\mathcal{A}^{0,1}(X) \longrightarrow \mathcal{A}^2(X))}, \quad H^q(X, \Omega_X) = 0 \ \text{for } q > 1.$$

Who is familiar with alternating differential forms in arbitrary dimensions will know the sequence

$$0 \longrightarrow \mathbb{R} \longrightarrow \mathcal{A}_X^0 \longrightarrow \mathcal{A}_X^1 \longrightarrow \cdots \longrightarrow \mathcal{A}_X^n \longrightarrow 0.$$

Here X is a differentiable manifold of dimension n and $\mathcal{A}_X^i$ denotes the sheaf of alternating differential forms of degree i. The general Lemma of Poincaré states that one gets an exact sequence if one takes global sections on an open subset which is diffeomorphic to an open convex domain in $\mathbb{R}^n$. Hence this sequence is an acyclic resolution of $\mathbb{R}_X$ for arbitrary X. We obtain the following general theorem.

§7. Differential forms

Theorem of de Rham. *For a differentiable manifold X one has*

$$H^i(X, \mathbb{C}_X) \cong \frac{\operatorname{Kern}(\mathcal{A}^i(X) \longrightarrow \mathcal{A}^{i+1}(X))}{\operatorname{Image}((\mathcal{A}^{i-1}(X) \longrightarrow \mathcal{A}^i(X))}.$$

The Stokes formula

We explain shortly the integration of differential forms. If $\omega = f_1 dx + f_2 dy$ is a 1-form on an open set $U \subset \mathbb{C}$ and if $\alpha : [0,1] \to U$ is a smooth (=infinitely differentiable) curve, $\alpha = \alpha_1 + i\alpha_2$, then one defines

$$\int_\alpha \omega := \int_0^1 \left(f_1(\alpha(t))\alpha_1'(t) + f_2(\alpha(t))\alpha_2'(t) \right) dt.$$

If $\omega = f dz$, then this integral coincides with the curve integral used in complex analysis. When $\gamma : U \to V$ is a differentiable map, then

$$\int_\alpha \gamma^* \omega = \int_{\gamma \circ \alpha} \omega.$$

This allows us to generalize this to Riemann surfaces. Let be $\omega \in \mathcal{A}^1(X)$ and let $\alpha : [0,1] \to X$ be a smooth curve. (It is clear how to define smoothness for α. One uses charts.) Then one can define $\int_\alpha \omega$. One divides α in pieces which lie in analytic charts. Then one uses the local formula. Because of the transformation invariance this is independent of the choice of the charts.

Similarly one defines the integral of a two-form ω. In the local case of an open subset $V \subset \mathbb{C}$ one takes for

$$\int_V f(z) dx \wedge dy$$

the usual 2-dimensional integral of f (if it exists). Again we need transformation invariance. If $\gamma : U \to V$ is a diffeomorphism, then

$$\int_V \omega = \int_U \gamma^* \omega$$

only holds if the (real) Jacobi determinant of γ is positive. When γ is biholomorphic, this is automatically the case, since the real Jacobi determinant of γ is $|\gamma'|^2$. This is an important fact because otherwise we could not integrate 2-forms on Riemann surfaces. We want to keep it in mind:

7.7 Proposition. *Riemann surfaces are **oriented** in the following sense. The chart change map of two analytic charts has positive real Jacobi determinant.*

If ω is a 2-form on X which vanishes outside a chart φ, then we can define

$$\int_X \omega := \int_{V_\varphi} \omega_\varphi.$$

The general case is reduced to this one by means of a partition of unity. Now we can formulate the formula of Stokes:

7.8 Theorem of Stokes. *Let $U \subset X$ be an open subset of a Riemann surface with compact closure. Assume that the boundary of U is the disjoint union of a finite number of (the images of) closed double point free regular curves $\alpha_i : [0,1] \to X$, $1 \le i \le n$. Assume furthermore that U is on the left of these curves. Then for every one-form ω on X*

$$\int_U d\omega = \sum_{i=1}^n \int_{\alpha_i} \omega.$$

We recall that a closed curve α is called double point free if $\alpha(t) = \alpha(t')$ holds only for $t, t' \in \{0, 1\}$. Regular means $\alpha'(t) \ne 0$ for all t in case of a plain curve $\alpha : [0, 1] \to \mathbb{C}$. The general case is reduced to that one by means of charts. An open subset $U \subset \mathbb{C}$ is called to be on the left of a regular curve α if for every t the following condition is satisfied. Consider the two vectors of norm 1 which are orthogonal to the tangent vector $\alpha'(t)$. Call $\mathbf{n}^+$ the one which goes to the right (this makes sense in the plane) and $\mathbf{n}^-$ the other one. Then there exists $\varepsilon > 0$ such that for all $0 \le s \le \varepsilon$

$$\alpha(t) + s\mathbf{n}^- \in U, \quad \alpha(t) + s\mathbf{n}^+ \notin U.$$

This can be generalized to open subsets U in a Riemann surface by means of analytic charts. One has to use Proposition 7.7.

The residue theorem on compact Riemann surfaces

Recall that a holomorphic differential (=1-form) on a Riemann surface is a collection $f_\varphi(z)dz$, where f_φ is a holomorphic function $f_\varphi : V_\varphi \to \mathbb{C}$ for every analytic chart (an atlas of analytic charts is enough) such that for two charts φ, ψ with chart change map γ the transformation formula

$$f_\psi(\gamma z) = \gamma'(z) f_\varphi(z)$$

holds. Instead of holomorphic functions one can take meromorphic functions. This leads to the notion of a *meromorphic differential* on a Riemann surface.

§7. Differential forms

We recall the following residue formula from complex calculus: Let $\gamma : U \to V$ be a biholomorphic map of open subsets $U, V \subset \mathbb{C}$. Let $f : V \to \mathbb{C}$ be a holomorphic function and $a \in U$. Then

$$\operatorname{Res}(f(w), \gamma(a)) = \operatorname{Res}(\gamma'(z)f(\gamma z), a).$$

It is an important fact that the factor $\gamma'(z)$ occurs. This means that it makes no sense to talk about residues of *meromorphic functions* on Riemann surfaces. But for a *meromorphic differential* ω the definition

$$\operatorname{Res}(\omega, a) := \operatorname{Res}(\omega_\varphi, \varphi(a)) \qquad (a \in U_\varphi)$$

makes sense since it is independent of the choice of a chart.

7.9 Residue theorem. *Let ω be a meromorphic differential on a compact Riemann surface. Then*

$$\sum_{a \in X} \operatorname{Res}(\omega, a) = 0.$$

The sum is of course a finite sum. For the proof we take for each pole a a small neighborhood $U(a)$ such that the closures of two different $U(a)$ are disjoint and such that the boundaries are nice (take disks with respect to charts). Then we apply the Stokes formula to the set

$$U = X - \bigcup_a \overline{U(a)}.$$

Since ω is holomorphic, we have $d\omega = 0$. It follows

$$\sum_a \int_{\partial U(a)} \omega = 0.$$

This proves the Residue Theorem 7.9. □

If f is a non-vanishing meromorphic function on a connected Riemann surface, then we can consider the meromorphic differential df/f. From the formula

$$\operatorname{Res}\left(\frac{df}{f}, a\right) = \operatorname{Ord}(f, a)$$

follows the following result.

7.10 Remark. *Let f be a non-constant meromorphic function on a connected compact Riemann surface. The number of poles and zeros – counted with multiplicity – is the same.*

This remark can also be proved without residue theorem. In Sect. 5 (Proposition 5.8) we have seen that this result has a merely topological background.

7.11 Remark. *Every holomorphic function on a connected compact Riemann surface is constant.*

This follows from Remark 7.10, but it follows also already from the maximum principle of complex calculus.

Chapter VI. The Riemann–Roch theorem

1. Generalities about vector bundles

A module M over a ring R is called finitely generated if there exists an integer $n \geq 0$ and a surjective R-linear map $R^n \to M$. This means that there exist elements $m_1, \ldots m_n$ such that every element of M can be written as linear combination of them. The elements $m_1, \ldots, m_n$ are called generators of M. The module M is called a finitely generated *free module* if there exists an integer $n \geq 0$ such that M is isomorphic to R^n. This means that there exists a basis, i.e. a system $m_1, \ldots, m_n$ of generators such that each element of M can be written as linear combination in a unique way.

Similarly, an $\mathcal{O}_X$-module $\mathcal{M}$ over a geometric structure $\mathcal{O}_X$, or more generally over an arbitrary sheaf of rings, is called finitely generated if there exists a surjective $\mathcal{O}_X$-linear map $\mathcal{O}_X^n \to \mathcal{M}$. Of course this is understood in the sheaf theoretic sense. We recall that an arbitrary $\mathcal{O}_X$-linear map $\mathcal{O}_X^n \to \mathcal{M}$ is given by a system $s_1, \ldots, s_n$ of global sections. The map $\mathcal{O}_X(U)^n \to \mathcal{M}(U)$ then is given by $(f_1, \ldots, f_n) \mapsto f_1 s_1|U + \cdots + f_n s_n|U$. If this map is surjective (in the sheaf theoretic sense), then we call $s_1, \ldots, s_n$ a system of generators. So generators of an $\mathcal{O}_X$-module are global sections $s_1, \ldots, s_n$ such that for each $a \in X$ the stalk $\mathcal{M}_a$ is generated as $\mathcal{O}_{X,a}$-module by the germs of the s_i. This means that for each section $s \in \mathcal{M}(U)$ in an open neighborhood of a there is a possibly smaller open neighborhood $a \in V \subset U$ such that $s|V$ is a linear combination of the $s_i|V$ with coefficients from $\mathcal{O}_X(V)$.

An $\mathcal{O}_X$-module $\mathcal{M}$ is called finitely generated and free if it is isomorphic to $\mathcal{O}_X^n$. This means that there exist global sections $s_1, \ldots, s_n$ such that for each open $U \subset X$ each element of $\mathcal{M}(U)$ can be written as linear combination of $s_1|U, \ldots, s_n|U$ in a unique way. We call these global sections then a basis of $\mathcal{M}$. A vector bundle on a geometric space is a locally finitely generated and free $\mathcal{O}_X$-module. We are only interested in the case where the geometric space is a Riemann surface.

1.1 Definition. *A **vector bundle** on a geometric space $(X, \mathcal{O}_X)$ is a locally finitely generated and free $\mathcal{O}_X$-module, i.e. a $\mathcal{O}_X$-module such that every point admits an open neighborhood U such that $\mathcal{M}|U$ is isomorphic to $(\mathcal{O}_X|U)^n$ for some integer $n \geq 0$. If n can be taken to be always 1, then $\mathcal{M}$ is called a **line bundle**.*

§1. Generalities about vector bundles

When n is the same for all U, for example when X is connected, we call n the rank of $\mathcal{M}$. Hence line bundles are vector bundles of rank 1.

We give a basic example for a line bundle. A *divisor* on a Riemann surface is a map $D: X \to \mathbb{Z}$ such that $D(a) = 0$ outside a discrete set. We are mainly interested in compact surfaces. Then this means that $D(a) = 0$ outside a finite set. We write D as formal linear combination

$$D = \sum_{a \in X} D(a)(a).$$

So (a) is the divisor that is one on a and zero outside a. The set of all divisors is an additive group $\mathrm{Div}(X)$ (componentwise addition). To every meromorphic function, which is not zero on any connected component, we associate the so-called *principal divisor* (f),

$$(f)(a) = \mathrm{Ord}(f, a).$$

From the formula $(fg) = (f) + (g)$ follows that the set of all principal divisors is a subgroup $H(X) \subset \mathrm{Div}(X)$. The elements of the factor group $\mathrm{Div}(X)/H(X)$ are called *divisor classes*. Let D, D' be two divisors on X. We write $D' \geq D$ if $D'(a) \geq D(a)$ for all a. Let f be a meromorphic function. The notation

$$(f)(a) \geq n \qquad (a \in X, \; n \in \mathbb{Z})$$

means that $\mathrm{Ord}(f, a) \geq n$ if f doesn't vanish in a neighborhood of a. Hence $(f)(a) \geq n$ is always true if $f = 0$ in a neighborhood of a. Now we associate to a divisor D on X the following sheaf.

$$\mathcal{O}_D(U) := \{\, f: U \to \bar{\mathbb{C}} \text{ meromorphic}, \; (f) \geq -D \,\}.$$

Clearly $\mathcal{O}_D$ is an $\mathcal{O}_X$-module. We claim that it is a line bundle: let U be a sufficiently small neighborhood of a given point. Then there exists a meromorphic function g whose divisor on U coincides with $-D$. One simply has to take U so small that it is in the domain of definition of a chart and that at most one $a \in U$ with $D(a) \neq 0$ exists. The map

$$\mathcal{O}_X(U) \xrightarrow{\sim} \mathcal{O}_D(U), \quad f \mapsto gf,$$

then is an isomorphism. Since U can be replaced by smaller open subsets, we obtain $\mathcal{O}_D|U \cong \mathcal{O}_X|U$. So we see that g is a basis of $\mathcal{O}_D|U$. Conversely, it is easy to see that any basis element of $\mathcal{O}_D|U$ is a meromorphic function g on U which fits to $-D$. This means $(g) = D|U$.

Two vector bundles are called isomorphic if they are isomorphic as $\mathcal{O}_X$-modules.

1.2 Proposition. *Let X be a Riemann surface. Two divisors D and D' are in the same divisor class if and only if the line bundles $\mathcal{O}_D$ and $\mathcal{O}_{D'}$ are isomorphic.*

Proof. Let $D' = D + (f)$. Then we get an isomorphism $\mathcal{O}_D \to \mathcal{O}_{D'}$ which sends a section g to fg.

We prove the converse, let $\mathcal{O}_D \to \mathcal{O}_{D'}$ be an isomorphism of $\mathcal{O}_X$-modules. Let $U \subset X$ be a small open subset such that $\mathcal{O}_D|U$ has a basis g. Let g' be its image in $\mathcal{O}_{D'}(U)$. This is a basis of $\mathcal{O}_{D'}|U$. We consider the meromorphic function g'/g. This element is independent of the choice of the basis element g. Hence the meromorphic functions g'/g glue to a global meromorphic function f on X. Obviously $D' = D + (f)$. □

If D is a divisor on a compact Riemann surface, then the degree

$$\deg(D) = \sum_{a \in X} D(a)$$

is defined. The rule $\deg(D + D') = \deg(D) + \deg(D')$ holds. As we know, the degree of a principal divisor is zero. Hence $\deg D$ depends only on the divisor class. This enables us to define the degree of a line bundle that is associated to a divisor.

1.3 Remark. *Let $\mathcal{L}$ be a line bundle on a compact Riemann surface. Assume that there exists a divisor D such that $\mathcal{L} \cong \mathcal{O}_D$. The **degree***

$$\deg \mathcal{L} := \deg D$$

does not depend on the choice of D.

Let $\mathcal{M}, \mathcal{N}$ be two $\mathcal{O}_X$-modules. We denote by $\operatorname{Hom}_{\mathcal{O}_X}(\mathcal{M}, \mathcal{N})$ the set of all $\mathcal{O}_X$-linear maps $\mathcal{M} \to \mathcal{N}$. This is an $\mathcal{O}_X(X)$-module. More generally, we can consider for every open $U \subset X$

$$U \longmapsto \operatorname{Hom}_{\mathcal{O}_X|U}(\mathcal{M}|U, \mathcal{N}|U).$$

It is clear that this is presheaf. It is easy to check that it is actually a sheaf and moreover an $\mathcal{O}_X$-module. We denote it by

$$\mathcal{H}om_{\mathcal{O}_X}(\mathcal{M}, \mathcal{N}).$$

We denote by $\mathcal{O}_X(U)^{p \times q}$ the set of all $p \times q$-matrices with entries from $\mathcal{O}_X(U)$. This is a free $\mathcal{O}_X(U)$-module. There is an obvious natural isomorphism

$$\mathcal{H}om_{\mathcal{O}_X}(\mathcal{O}_X^p, \mathcal{O}_X^q) \cong \mathcal{O}_X^{p \times q}.$$

Hence $\mathcal{H}om_{\mathcal{O}_X}(\mathcal{M}, \mathcal{N})$ is a vector bundle if $\mathcal{M}$ and $\mathcal{N}$ are vector bundles. It is a line bundle if both are line bundles. The *dual bundle* of a vector bundle $\mathcal{M}$ is defined as

$$\mathcal{M}^* := \mathcal{H}om_{\mathcal{O}_X}(\mathcal{M}, \mathcal{O}_X).$$

§1. Generalities about vector bundles

It has the same rank as $\mathcal{M}$.

There is another construction which rests on the tensor product of modules. Let $\mathcal{M}, \mathcal{N}$ be two $\mathcal{O}_X$-modules. The assignment

$$U \longmapsto \mathcal{M}(U) \otimes_{\mathcal{O}_X(U)} \mathcal{N}(U)$$

defines a presheaf. This is usually not a sheaf. Hence we consider the generated sheaf and denote it by $\mathcal{M} \otimes_{\mathcal{O}_X} \mathcal{N}$. Clearly this is an $\mathcal{O}_X$-module. The notion of an $\mathcal{O}_X$-bilinear map $\mathcal{M} \times \mathcal{N} \to \mathcal{P}$ for $\mathcal{O}_X$-modules $\mathcal{M}, \mathcal{N}, \mathcal{P}$ and the following universal property should be clear. For an $\mathcal{O}_X$-bilinear map $\mathcal{M} \times \mathcal{N} \to \mathcal{P}$ of $\mathcal{O}_X$-modules there exists a unique commutative diagram

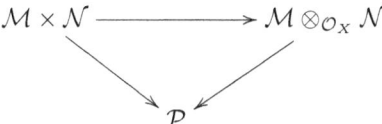

with an $\mathcal{O}_X$-linear map $\mathcal{M} \otimes_{\mathcal{O}_X} \mathcal{N} \to \mathcal{P}$.

One also has $\mathcal{O}_X^n \otimes_{\mathcal{O}_X} \mathcal{O}_X^m \cong \mathcal{O}_X^{n \times m}$. The construction of the tensor product is compatible with the restriction to open subsets,

$$(\mathcal{M} \otimes_{\mathcal{O}_X} \mathcal{N})|U \cong \mathcal{M}|U \otimes_{\mathcal{O}_X|U} \mathcal{N}|U.$$

This follows from the fact the construction of the generated sheaf $\hat{F}$ is compatible with restriction to open subsets. We obtain that the tensor product of two vector bundles is a vector bundle and moreover, the tensor product of two line bundles is a line bundle.

When M is an R-module and $M^* = \mathrm{Hom}_R(M, R)$ its dual, one has a natural bilinear map $M \times M^* \longrightarrow R$, $(a, l) \mapsto l(a)$. This induces a linear map $M \otimes_R M^* \to R$. Sheafifying we get an $\mathcal{O}_X$-bilinear map

$$\mathcal{M} \otimes_{\mathcal{O}_X} \mathcal{M}^* \longrightarrow \mathcal{O}_X.$$

Clearly, this is an isomorphism when $\mathcal{M}$ is a line-bundle.

We denote by $[\mathcal{L}]$ the class of all line bundles which are isomorphic to $\mathcal{L}$. We denote be $\mathrm{Pic}(X)$ (Picard group) the set of all isomorphy classes of line-bundles.

Since constructions as the "set of sets" are forbidden in set theory, there seems to be a logical difficulty. But this is not really there because one can easily prove that there exists a *set* of line-bundles such that every line bundle is isomorphic to one of this set.

We define a product in $\mathrm{Pic}(X)$ by

$$[\mathcal{L}] \otimes [\mathcal{L}'] := [\mathcal{L} \otimes_{\mathcal{O}_X} \mathcal{L}'].$$

This is clearly well-defined. We notice that this product is associative and commutative.

The element $[\mathcal{O}_X]$ is a neutral element in $\mathrm{Pic}(X)$. Finally we notice that $[\mathcal{L}] \otimes [\mathcal{L}^*] = [\mathcal{O}_X]$. Hence every element of $\mathrm{Pic}(X)$ has an inverse.

1.4 Remark. *The set $\mathrm{Pic}(X)$ of isomorphy classes of line bundles is an abelian group under the tensor product. The trivial bundle $\mathcal{O}_X$ gives the unit element, the dual of a line bundle gives the inverse.*

This group is called the *Picard group*.

2. The finiteness theorem

We introduce the notion of a coherent sheaf.

2.1 Definition. *An $\mathcal{O}_X$-module $\mathcal{M}$ on a Riemann surface X is called a skyscraper sheaf if $\mathcal{M}_a$ is zero for all a outside a discrete set S and if $\mathcal{M}_a$ is a finite dimensional complex vector space for all $a \in S$.*

The skyscraper sheaf is determined by the stalks:

2.2 Remark. *For a skyscraper sheaf the natural map*

$$\mathcal{M}(U) \xrightarrow{\sim} \prod_{a \in U} \mathcal{M}_a \quad (U \subset X \text{ open})$$

is an isomorphism.

Proof. The map is injective because $\mathcal{M}$ is a sheaf. To prove the surjectivity, we consider the discrete set $S \subset U$ of all a with $\mathcal{M}_a \neq 0$. Let $(s_a) \in \prod \mathcal{M}_a$. We choose for each $a \in S$ an open neighborhood $U(a) \subset U$ such that $U(a) \cap S$ is $\{a\}$ and such that s_a can be represented by a section $s_{U(a)} \in \mathcal{M}(U(a))$. These sections and the zero section on $U - S$ glue to a section on U. □

The notion of a "coherent sheaf" includes skyscraper sheafs and vector bundles.

2.3 Definition. *A **coherent sheaf** on a Riemann surface is an $\mathcal{O}_X$-module $\mathcal{M}$ such that there exists an exact sequence*

$$0 \longrightarrow \mathcal{W} \longrightarrow \mathcal{M} \longrightarrow \mathcal{N} \longrightarrow 0$$

with a skyscraper sheaf $\mathcal{W}$ and a vector bundle $\mathcal{N}$.

An element r of a ring R is called a *zero divisor* if multiplication by r is not an injective map $R \to R$. An element a of an R-module M is called a *torsion element* if there exists a non-zero divisor $r \in R$ such that $ra = 0$. The set of all torsion elements is the torsion submodule M^{tor}.

§2. The finiteness theorem

2.4 Lemma. *Let*
$$0 \longrightarrow \mathcal{W} \longrightarrow \mathcal{M} \longrightarrow \mathcal{N} \longrightarrow 0$$
be an exact sequence as in Definition 2.3. Then $\mathcal{W}_a$ is the torsion sub-module of $\mathcal{M}_a$.

Proof. It is helpful to make use of some very simple structure theorem of algebra. Recall that an R-submodule $\mathfrak{a}$ of a ring R is called an *ideal* in R. If a is an element of R then Ra is an ideal and such an ideal is called a *principal ideal*. A ring is called *principal ideal ring* if every ideal is a principal ideal.

We are interested in the ring $\mathbb{C}\{z\}$ of all convergent power series in one variable. If a is a point of a Riemann surface, then $\mathcal{O}_{X,a}$ is isomorphic to $\mathbb{C}\{z\}$. From the point of view of algebra, the ring $\mathbb{C}\{z\}$ is an extremely simple ring. Every ideal is a principal ideal which is generated by z^n for a suitable integer $n \geq 0$. A known result of algebra states that every finitely generated module M over a principal ideal ring R is isomorphic to the finite direct product of modules of the type $R/\mathfrak{a}$. In the case $R = \mathbb{C}\{z\}$ the proof is very simple.

Since $\mathcal{W}_a$ is a finite dimensional vector space, the structure theorem shows that all its elements are torsion elements. Hence it is contained in $\mathcal{M}_a^{\text{tor}}$. Conversely every torsion element of $\mathcal{M}_a$ maps to 0, since $\mathcal{N}_a$ is torsion free. This proves Lemma 2.4. □

Lemma 2.4 implies that "coherence" is a local property:

Let (U_i) be an open covering. An $\mathcal{O}_X$-module $\mathcal{M}$ is coherent if and only if all $\mathcal{M}|U_i$ are coherent.

2.5 Theorem. *Let $\mathcal{M}$ be a coherent sheaf on a compact Riemann surface. Then the following holds:*
a) $H^n(X, \mathcal{M})$ *is finite dimensional for all n.*
b) $H^n(X, \mathcal{M}) = 0$ *for $n > 1$.*

Proof. Skyscraper sheaves are flabby. Hence all higher cohomology groups vanish. The long exact cohomology sequence shows that we can restrict to vector bundles.

First we show the vanishing statement b). We will use the Dolbeault complex $0 \to \mathcal{O}_X \to \mathcal{A}_X^0 \to \mathcal{A}_X^{0,1} \to 0$. We know that this is an exact sequence (Lemma of Dolbeault V.7.3). The essential point is that $\bar{\partial}: \mathcal{A}_X^0 \to \mathcal{A}_X^{0,1}$ is $\mathcal{O}_X$-linear. This comes from the fact that $\bar{\partial} f = 0$ for holomorphic functions. Hence we can consider the Dolbeault complex as a sequence of $\mathcal{O}_X$-modules and we can tensor it with $\mathcal{M}$. The sequence

$$0 \longrightarrow \mathcal{M} \longrightarrow \mathcal{A}_X^0 \otimes_{\mathcal{O}_X} \mathcal{M} \to \mathcal{A}^{0,1} \otimes_{\mathcal{O}_X} \mathcal{M} \to 0$$

remains of course exact. Since $\mathcal{A}_X^\cdot \otimes_{\mathcal{O}_X} \mathcal{M}$ carries a natural structure as $\mathcal{C}_X^\infty$-module, we have an acyclic resolution of $\mathcal{M}$ (Proposition V.3.2). The long exact cohomology sequence gives $H^i(X, \mathcal{M}) = 0$ for $i > 1$.

Next we prove the finiteness of $H^0(X, \mathcal{M})$. We choose a finite open covering $\mathfrak{U} = (U_i)$ such that each U_i is biholomorphic equivalent to a disk in the plane. This is possible since X is compact. We can take the U_i so small that $\mathcal{M}|U_i$ is free. We choose a trivialization $\mathcal{M}|U_i \cong \mathcal{O}_{U_i}^n$. We choose open subsets $V_i \subset U_i$ whose closure (taken in X) is contained in U_i and such that $\mathcal{V} = (V_i)$ is still a covering of X. We can assume that all V_i are biholomorphic equivalent to disks. By means of the isomorphism $\mathcal{M}(U_i) \cong \mathcal{O}(U_i)^n$ we can equip $\mathcal{M}(U_i)$ with a structure as a Frèchet space. Here we use Remark I.7.4 and the trivial fact that a finite direct product of Frèchet spaces is a Frèchet space (Remark I.7.2). In particular, $\prod_i \mathcal{M}(U_i)$ also carries a structure as Frèchet space. We can consider $\mathcal{M}(X)$ as a (closed) subspace of this space. A closed subspace of a Frèchet space is also a Frèchet space (Remark I.7.2). Hence $\mathcal{M}(X)$ is a Frèchet space too. We can repeat this with the covering $\mathfrak{V}$. It is easy to see that both Frèchet structures on $\mathcal{M}(X)$ are the same. From Montel's theorem (in the form of Proposition I.7.8) follows that the identity map $\mathcal{M}(X) \to \mathcal{M}(X)$ is compact. From Corollary I.7.7 follows that $\mathcal{M}(X)$ is finite dimensional.

Next we prove the finiteness of $H^1(X, \mathcal{M})$. From Leray's lemma IV.3.6 we know

$$H^1(X, \mathcal{M}) = \check{H}^1(\mathfrak{U}, \mathcal{M}) = \check{H}^1(\mathfrak{V}, \mathcal{M}).$$

We consider the space of Čech one-cocycles $C^1(\mathfrak{U}, \mathcal{M})$. This is a closed subspace of $\prod_{ij} \mathcal{M}(U_i \cap U_j)$ and hence a Frèchet space. The same is true for $C^1(\mathfrak{V}, \mathcal{M})$. The natural restriction operator

$$r : C^1(\mathfrak{U}, \mathcal{M}) \longrightarrow C^1(\mathfrak{V}, \mathcal{M})$$

is a compact operator by Montel's theorem (in the form of Proposition I.7.8). Now we consider the map

$$C^1(\mathfrak{U}, \mathcal{M}) \times \prod_i \mathcal{M}(V_i) \longrightarrow C^1(\mathfrak{V}, \mathcal{M}),$$

$$(A, (s_i)) \longmapsto r(A) + \delta((s_i)).$$

This map is a continuous surjective linear map of Frèchet spaces. It differs from the map

$$C^1(\mathfrak{U}, \mathcal{M}) \times \prod_i \mathcal{M}(V_i) \longrightarrow C^1(\mathfrak{V}, \mathcal{M}),$$

$$(A, (s_i)) \longmapsto \delta((s_i)).$$

only by a compact operator (essentially r). By the Schwartz theorem I.7.6 this map has a finite dimensional cokernel. But the cokernel is $\check{H}^1(\mathfrak{V}, \mathcal{M})$. □

3. The Picard group

In Sect. 1 we introduced the Picard group $\mathrm{Pic}(X)$, i.e. the group of isomorphism classes of line bundles. We want to compare it with the group $\mathrm{Div}(X)/H(X)$ of divisor classes. Recall that we associated to a divisor D on a Riemann surface X a line bundle $\mathcal{O}_D$. For two divisors D, D' there is a natural $\mathcal{O}_X$-bilinear multiplication map

$$\mathcal{O}_D \times \mathcal{O}_{D'} \longrightarrow \mathcal{O}_{D+D'}.$$

It is clear that the induced homomorphism

$$\mathcal{O}_D \otimes_{\mathcal{O}_X} \mathcal{O}_{D'} \longrightarrow \mathcal{O}_{D+D'}$$

is an isomorphism. Hence we get a *homomorphism* $\mathrm{Div}(X) \to \mathrm{Pic}(X)$. Since principal divisors are in the kernel (Proposition 1.2) we get a homomorphism of the group of divisor classes into $\mathrm{Pic}(X)$. This homomorphism is injective (Proposition 1.2). We claim more:

3.1 Proposition. *Let X be a compact Riemann surface. The natural homomorphism*

$$\mathrm{Div}(X)/H(X) \xrightarrow{\sim} \mathrm{Pic}(X)$$

is an isomorphism.

It remains to show that every line bundle is isomorphic to the line bundle of a divisor. The proof is non-trivial and will use the finiteness theorem. We will use twists of sheaves. The twist of a coherent sheaf $\mathcal{M}$ with a line bundle $\mathcal{L}$ simply is $\mathcal{M} \otimes_{\mathcal{O}_X} \mathcal{L}$. We will use a very special twist. For the rest of the section we fix a point $a \in X$. For an integer n we consider the divisor $n(a)$

$$n(a)(x) = \begin{cases} n & \text{for } x = a, \\ 0 & \text{elsewhere.} \end{cases}$$

We denote by $\mathcal{O}(n)$ the associated line bundle and by $\mathcal{M}(n)$ the twist of $\mathcal{M}$ with $\mathcal{O}(n)$. (This construction depends on the choice of a.)

We assume now for simplicity that $\mathcal{M}$ is a vector bundle. We use the natural inclusion $\mathcal{O}(n) \hookrightarrow \mathcal{O}(n+1)$. It induces $\mathcal{M}(n) \hookrightarrow \mathcal{M}(n+1)$. We obtain an exact sequence

$$0 \longrightarrow \mathcal{M}(n) \longrightarrow \mathcal{M}(n+1) \longrightarrow \mathcal{W}(n) \longrightarrow 0$$

with a skyscraper sheaf $\mathcal{W}(n)$. From the exact cohomology sequence we get that $H^1(X, \mathcal{M}(n)) \to H^1(X, \mathcal{M}(n+1))$ is surjective. By the finiteness theorem these are finite dimensional vector spaces. The dimension d_n of $H^1(X, \mathcal{M}(n))$ gives a decreasing sequence of non-negative integers. This must be constant for big n. Hence we obtain the following result.

3.2 Lemma. *Let $\mathcal{M}$ be a vector bundle on a compact Riemann surface X. For sufficiently large n the homomorphism*
$$H^1(X, \mathcal{M}(n)) \longrightarrow H^1(X, \mathcal{M}(n+1))$$
is an isomorphism.

Now we make use of the exact sequence
$$H^0(X, \mathcal{M}(n+1)) \longrightarrow H^0(X, \mathcal{W}(n)) \longrightarrow H^1(X, \mathcal{M}(n)) \longrightarrow H^1(X, \mathcal{M}(n+1)).$$

For large n the last arrow is an isomorphism, hence $H^0(X, \mathcal{M}(n+1)) \to H^0(X, \mathcal{W}(n))$ is surjective. Since $H^0(X, \mathcal{W}(n))$ is not zero, we obtain the following proposition.

3.3 Proposition. *Let $\mathcal{M}$ be a vector bundle of positive rank on a compact Riemann surface. Then $\mathcal{M}(n)$ admits for large enough n a non-zero global section.*

Taking products, one gets a bilinear map $\mathcal{O}(-n) \times \mathcal{O}(n) \to \mathcal{O}_X$. More generally, one gets for a vector bundle $\mathcal{M}$ a natural map
$$\mathcal{M} \times \mathcal{O}(-n) \times \mathcal{M}^* \times \mathcal{O}(n) \longrightarrow \mathcal{O}_X.$$
This induces an $\mathcal{O}_X$-linear map
$$\mathcal{M}(-n)^* \xrightarrow{\sim} \mathcal{M}^*(n).$$

A local computation shows that this is an isomorphism. For suitable (large enough) n one has a non-zero global section $S \in H^0(X, \mathcal{M}^*(n))$. We can use this section to define a non-zero $\mathcal{O}_X$-linear map
$$\mathcal{M}(-n) \longrightarrow \mathcal{O}_X, \quad (a \longmapsto S(a)).$$

Next we need a lemma.

3.4 Lemma. *Let $\mathcal{M}$ be a vector bundle on a Riemann surface and let $\mathcal{M} \to \mathcal{O}_X$ be a non-zero $\mathcal{O}_X$-linear map. Then the image is a line bundle of the form $\mathcal{O}_{-D}$ with $D \geq 0$.*

Proof. We can replace X by a small open neighborhood of a given point a. Hence we may assume that $X = U$ is an open neighborhood of $a = 0$ in $\mathbb{C}$. Taking U small enough, we can assume $\mathcal{M} = \mathcal{O}_U^d$. The map $\mathcal{M} \to \mathcal{O}_U$ is given by a system of holomorphic functions $f_1, \ldots, f_d \in \mathcal{O}(U)$. We can assume that none of the f_i is zero and then that $f_i = z^{n_i} g_i$ where g_i is without zero. Modifying the map $\mathcal{M} \to \mathcal{O}_U$ we reduce to the case $g_i = 1$. Now we see that the image is $z^n \mathcal{O}_U$, where n is the minimum of the $n_1, \ldots, n_d$. This proves Lemma 3.4. □

The same type of argument gives also information about the kernel. We can assume $n_1 \leq \ldots \leq n_d$. But then the first component of a section of the kernel can be computed from the rest. Hence we get that the kernel again is free. This shows the following important lemma.

§4. Riemann–Roch

3.5 Coherence lemma. *Let $\mathcal{M} \to \mathcal{L}$ be an $\mathcal{O}_X$-linear map of a vector bundle into a line bundle (on a Riemann surface). Then the kernel is a vector bundle too.*

We go back to the map $\mathcal{M}(n) \to \mathcal{O}_X$. We know that the image is a line bundle. Tensoring with $\mathcal{O}(-n)$ we obtain the following result.

3.6 Lemma. *Every vector bundle $\mathcal{M}$ of rank $n > 0$ sits in an exact sequence $0 \to \mathcal{N} \to \mathcal{M} \to \mathcal{L} \to 0$, where $\mathcal{L}$ is a line bundle, and where $\mathcal{N}$ is a vector bundle of rank $n - 1$. The line bundle $\mathcal{L}$ can be taken in the form $\mathcal{O}_D$ for some divisor D.*

Since a vector bundle of rank 0 is zero, we obtain the proof of Proposition 3.1. □

When we apply Proposition 3.3 to $\mathcal{M} = \mathcal{O}_X$, we get the following theorem.

3.7 Theorem. *Every compact Riemann surface admits a non-constant meromorphic function.*

4. Riemann–Roch

In the following X denotes a connected Riemann surface. For a vector bundle on a connected Riemann surface we have defined the rank $n = \text{rank}(\mathcal{M})$. This can be generalized to a coherent sheaf $\mathcal{M}$: there exists a discrete subset $S \subset X$ such that $\mathcal{M}|(X - S)$ is a vector bundle. We define $\text{rank}(\mathcal{M}) := \text{rank}(\mathcal{M}|(X - S))$. The rank of a skyscraper sheaf is 0.

4.1 Remark. *For a short exact sequence of coherent sheaves*
$$0 \longrightarrow \mathcal{M}_1 \longrightarrow \mathcal{M}_2 \longrightarrow \mathcal{M}_3 \longrightarrow 0$$
one has
$$\text{rank}(\mathcal{M}_2) = \text{rank}(\mathcal{M}_1) + \text{rank}(\mathcal{M}_3).$$

Proof. One has to use the following. Assume that R is an integral domain and
$$0 \longrightarrow R^m \longrightarrow R^n \longrightarrow R^p \longrightarrow 0$$
an exact sequence of R-modules. Then $n = m + p$. This is well-known from linear algebra when R is a field and can be reduced to this case by imbedding R into a field. □

There are other quantities which have the additive property as in Remark 4.1. For a coherent sheaf we define
$$\chi(\mathcal{M}) = \sum_i (-1)^i \dim H^i(X, \mathcal{M}) = \dim H^0(X, \mathcal{M}) - \dim H^1(X, \mathcal{M}).$$
In analogy to Remark 4.1 we have the following result.

4.2 Remark. *For a short exact sequence of coherent sheaves*

$$0 \longrightarrow \mathcal{M}_1 \longrightarrow \mathcal{M}_2 \longrightarrow \mathcal{M}_3 \longrightarrow 0$$

one has

$$\chi(\mathcal{M}_2) = \chi(\mathcal{M}_1) + \chi(\mathcal{M}_3).$$

Proof. One has to use the long exact cohomology sequence and the following simple result of linear algebra: let

$$0 \longrightarrow V_1 \longrightarrow \cdots \longrightarrow V_n \longrightarrow 0$$

be an exact sequence of finite dimensional vector spaces, then

$$\sum_{i=1}^{n}(-1)^i \dim V_i = 0. \qquad \square$$

4.3 Theorem. *Let X be a compact Riemann surface. There exists a unique function which associates to an arbitrary coherent sheaf $\mathcal{M}$ on X a non-negative integer $\deg \mathcal{M}$ such that the following properties are satisfied:*
1) $\deg \mathcal{M}$ *depends only on the isomorphy class of* $\mathcal{M}$.
2) *For a skyscraper sheaf* $\mathcal{W}$

$$\deg(\mathcal{W}) = \sum_{a \in X} \dim \mathcal{W}_a.$$

3) *If D is a divisor and $\mathcal{O}_D$ the associated line-bundle, then*

$$\deg(\mathcal{O}_D) = \deg D.$$

4. *For a short exact sequence of coherent sheaves*

$$0 \longrightarrow \mathcal{M}_1 \longrightarrow \mathcal{M}_2 \longrightarrow \mathcal{M}_3 \longrightarrow 0$$

one has

$$\deg(\mathcal{M}_2) = \deg(\mathcal{M}_1) + \deg(\mathcal{M}_3).$$

We will prove this together with the Riemann–Roch theorem. In this theorem a fundamental invariant occurs, the *genus* of the Riemann surface. It is defined (for compact Riemann surfaces) by

$$\boxed{g := \dim H^1(X, \mathcal{O}_X)}$$

§4. Riemann–Roch

We shall give different characterizations of the genus later. Using *Serre duality*, we will see in Theorem 7.2 the formula $g = \dim \Omega(X)$. The theorem of de Rham–Hodge VII.2.5 implies $\dim H^1(X, \mathbb{C}) = 2g$. In particular, the genus is a pure topological invariant. A famous topological result states that two compact Riemann surfaces with the same genus are homeomorphic. They look like a sphere with g handles. Another way to express this is that in the case $g \geq 2$ it is obtained from g tori by connecting them with tunnels. Here is a picture for $g = 2$.

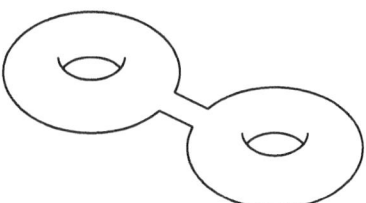

Proofs can be found for example in [Fr1].

4.4 Riemann–Roch theorem. *Let $\mathcal{M}$ be a coherent sheaf on a compact Riemann surface X. Then*

$$\chi(\mathcal{M}) = \deg(\mathcal{M}) + \operatorname{rank}(\mathcal{M})(1 - g).$$

Corollary. $\dim H^0(X, \mathcal{M}) \geq \deg(\mathcal{M}) + \operatorname{rank}(\mathcal{M})(1 - g)$.

So far, the degree has been defined for skyscraper sheaves (Theorem 4.3, 2)) and for line bundles (Theorem 4.3, 3)). The Riemann–Roch theorem is trivial for skyscraper sheaves. Since the rank of a skyscraper sheaf is 0 it reduces to the equation $\dim \mathcal{M}(X) = \deg \mathcal{M}$ which follows from Remark 2.2.

The Riemann–Roch theorem is also trivial for $\mathcal{O}_X$, since

$$\chi(\mathcal{O}_X) = \dim H^0(X, \mathcal{O}_X) - \dim H^1(\mathcal{O}_X) = 1 - g.$$

Next we prove the Riemann–Roch theorem for a line bundle. We know that it is isomorphic to the line bundle associated to a divisor D. We add to the divisor D a point divisor (a). We have an exact sequence

$$0 \longrightarrow \mathcal{O}_D \longrightarrow \mathcal{O}_{D+(a)} \longrightarrow \mathcal{W} \longrightarrow 0$$

with a skyscraper sheaf $\mathcal{W}$. Since we have $\deg(D + (a)) = \deg(D) + \deg(\mathcal{W})$, the Riemann–Roch theorem holds for $\mathcal{O}_D$ if and only if it holds for $\mathcal{O}_{D+(a)}$. Adding and subtracting point divisors, one can reduce the case of an arbitray D to the zero divisor in a finite number of steps. But for the zero divisor ($\mathcal{L} = \mathcal{O}_X$) Riemann–Roch has been proved.

We now make a little trick. We *define* $\deg(\mathcal{M})$ in general by the Riemann–Roch formula. For skyscraper sheaves and for line bundles this coincides with the definition we gave already. But also 4) in Theorem 4.3 is true since additivity holds for χ and rank. The uniqueness of deg with the properties 1)-4) follows from Proposition 3.1 and Lemma 3.6. This proves the existence of the degree and the Riemann–Roch theorem. □

4.5 Lemma. *Let $\mathcal{M}, \mathcal{N}$ be two vector bundles, then*
$$\deg(\mathcal{M} \otimes_{\mathcal{O}_X} \mathcal{N}) = \deg \mathcal{M} \cdot \mathrm{rank} \mathcal{N} + \deg \mathcal{N} \cdot \mathrm{rank} \mathcal{M}.$$
Let $\mathcal{M}, \mathcal{N}$ be two vector bundles than
$$\deg(\quad \mathcal{O}_X(\mathcal{M}, \mathcal{N})) = \deg \mathcal{N} - \deg \mathcal{M}.$$

Proof. For line bundles the statements are clear. The general case can be reduced to this one by induction on the ranks using exact sequences of the type as $0 \to \mathcal{M}_1 \to \mathcal{M} \to \mathcal{L} \to 0$ where $\mathcal{L}$ is a line bundle (Lemma 3.6) and using the exact sequences Remark II.3.1 and Remark II.3.2. □

There is another useful formula. Since we will use it only in the Appendix (Chapt VIII, Sects 7 and 8), we keep short:

Let M be a module over a ring R. One can consider for any natural number n the n-fold tensor product
$$M^{\otimes n} = \overbrace{M \otimes \cdots \otimes M}^{n\text{-times}}.$$
There is a natural multilinear mapping
$$\overbrace{M \times \cdots \times M}^{n\text{-times}} \longrightarrow M^{\otimes n}, \quad (m_1, \ldots, m_n) \longmapsto m_1 \otimes \cdots \otimes m_n,$$
which is universal in the following sense. For any multilinear map from $M \times \cdots \times M$ into some R-module N there exists a linear map which makes the following diagram commutative.

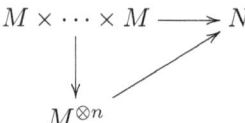

We consider the submodule J of $M^{\otimes n}$ which is generated by all elements of the form $m_1 \otimes \cdots \otimes m_n$ such that at least two of the m_i are equal. Then we define the *exterior power*
$$\bigwedge^n M = M^{\otimes n}/J.$$
Let $m_1, \ldots, m_n \in M \times \cdots \times M$. We denote by $m_1 \wedge \ldots \wedge m_n$ the coset of $m_1 \otimes \cdots \otimes m_n$. We obtain an alternating multilinear map
$$M \times \cdots \times M \longrightarrow \bigwedge^n M, \quad (m_1, \ldots, m_n) \longmapsto m_1 \wedge \ldots \wedge m_n.$$
This has a universal property with respect to alternating multilinear maps of M^n into an arbitrary module N. Its formulation and proof can be left to the reader. We need a very particular case. Let M be free of rank n and let $e_1, \ldots, e_n$ be a basis of M. Then $\bigwedge^n M$ is free of rank one. Actually $e_1 \wedge \ldots \wedge e_n$ is a basis element.

§4. Riemann–Roch

Let $\mathcal{M}$ be an $\mathcal{O}_X$-module. One can consider the presheaf

$$U \longmapsto \bigwedge^n \mathcal{M}(U)$$

with obvious restriction maps. We denote by $\bigwedge^n \mathcal{M}$ the generated sheaf which is an $\mathcal{O}_X$-module. This construction is compatible with the restriction to open subsets (since the tensor product and the construction "generated sheaf" have this property).

Let $\mathcal{M} = \mathcal{O}_X^n$. Then the presheaf $U \mapsto \bigwedge^n(\mathcal{M}(U)) \cong \mathcal{O}_X(U)$ is already a sheaf and this sheaf is isomorphic to $\mathcal{O}_X$. This shows that $\bigwedge^n \mathcal{M}$ is a line bundle if $\mathcal{M}$ is a vector bundle of rank n. This line bundle is called the determinant of $\mathcal{M}$ and is denoted by $\det \mathcal{M}$. If $U \subset X$ is small enough (such that the restriction $\mathcal{M}|U$ is trivial), then $(\bigwedge^n \mathcal{M})(U) = \bigwedge^n(\mathcal{M}(U)) \cong \mathcal{O}_X(U)$.

4.6 Remark. *A vector bundle and its determinant have the same degree.*

Proof. Let $0 \to \mathcal{M}_1 \to \mathcal{M}_2 \to \mathcal{M}_3 \to 0$ be a short exact sequence and let p, q be two natural numbers. Set $n = p + q$. Assume that

$$a_1 \wedge \ldots \wedge a_m = 0 \quad \text{for} \quad a_i \in \mathcal{M}_1 \quad \text{if} \quad m > p.$$

This is the case if M_1 can be generated by p elements. Then the map

$$M_1^p \times M_3^q \longrightarrow \bigwedge^n M_2, \quad (a_1, \ldots, a_p, b_1, \ldots b_q) \longmapsto a_1 \wedge \ldots \wedge a_p \wedge B_1 \wedge \ldots \wedge B_q,$$

where $B_i \in M_2$ is an inverse image of b_2, is well defined. Using the universal property of the tensor resp. the alternating product, we conclude that this map factors through

$$\bigwedge^p M_1 \otimes \bigwedge^q M_3 \longrightarrow \bigwedge^n M_2.$$

Assume now that $M_1 \cong R^p$, $M_2 \cong R^n$, $M_3 \cong R^q$. Then this map is an isomorphism.

Let now $0 \to \mathcal{M}_1 \to \mathcal{M}_2 \to \mathcal{M}_3 \to 0$ be a short exact sequence of vector bundles, $\mathcal{M}_1$ of rank p, $\mathcal{M}_3$ of rank q and $\mathcal{M}_2$ of rank $n = p + q$.

Then the above consideration gives an $\mathcal{O}_X$-linear map

$$\bigwedge^p \mathcal{M}_1 \otimes \bigwedge^q \mathcal{M}_3 \xrightarrow{\sim} \bigwedge^n \mathcal{M}_2$$

where a local consideration shows that it is an isomorphism. This formula shows $\deg \det \mathcal{M}_2 = \deg \det \mathcal{M}_1 + \deg \det \mathcal{M}_3$. Recall that we also have $\deg \mathcal{M}_2 = \deg \mathcal{M}_1 + \deg \mathcal{M}_2$. Hence Remark 4.6 is true for $\mathcal{M}_2$ if it is true for $\mathcal{M}_1$ and $\mathcal{M}_3$. Now we can prove Remark 4.6 by induction on the rank. The beginning of the induction is trivial, since the determinant of a line bundle is the line bundle itself. The induction step uses short exact sequences as in Lemma 3.6. □

5. A residue map

In this section X is a compact (and connected) Riemann surface. We now have to make use of the sheaf Ω_X of holomorphic differentials. For sake of simplicity we shall write $\mathcal{O} = \mathcal{O}_X$, $\Omega = \Omega_X$. Basic will be the „residue map"

$$\mathrm{Res} : H^1(X, \Omega) \longrightarrow \mathbb{C}.$$

5.1 Remark and first definition of the residue map. *Consider the exact sequence*

$$0 \longrightarrow \Omega \longrightarrow \mathcal{A}^{1,0} \xrightarrow{\bar{\partial}} \mathcal{A}^2 \longrightarrow 0$$

and the resulting isomorphism

$$H^1(X, \Omega) \cong \mathcal{A}^2(X)/\bar{\partial}\mathcal{A}^{1,0}.$$

For an $\xi \in H^1(X, \Omega)$ consider a representative $\omega \in \mathcal{A}^2(X)$. The integral

$$\mathrm{Res}\,\xi := \frac{1}{2\pi i} \int_X \omega$$

is independent of the choice of the representative ω.

The independence follows from Stokes' theorem $\int_X d\eta = 0$. Notice that on the level of $\mathcal{A}^{1,0}$ we have $\bar{\partial} = d$. □

We come to another property which justifies the notation "residue". For this we consider the sheaf $\mathfrak{M}$ of meromorphic 1-forms and the exact sequence

$$0 \longrightarrow \Omega \longrightarrow \mathfrak{M} \longrightarrow \mathfrak{M}/\Omega \longrightarrow 0.$$

This induces the map

$$H^0(X, \mathfrak{M}/\Omega) \xrightarrow{\delta} H^1(X, \Omega).$$

§5. A residue map

We define now a residue map

$$\mathrm{Res}: H^0(X, \mathfrak{M}/\Omega) \longrightarrow \mathbb{C}.$$

Recall that the elements of $(\mathfrak{M}/\Omega)(X)$ are families $(\omega_a)_{a\in X}$, $\omega \in \mathcal{M}_a/\mathcal{O}_a$, which locally fit together. This means that there exists an open covering $\mathfrak{U} = (U_i)$ and meromorphic differentials $\omega_i \in \mathfrak{M}(U_i)$, which represent the ω_a for all $a \in U_i$. The differences $\omega_i - \omega_j$ are holomorphic in $U_i \cap U_j$. Hence the residue $\mathrm{Res}(\omega_a, a)$ is well defined.

5.2 Remark and second definition of the residue. *There is a natural linear map*

$$\mathrm{Res}: H^0(X, \mathfrak{M}/\Omega) \longrightarrow \mathbb{C},$$

which can be defined as

$$\mathrm{Res}((\omega_a)) = \sum_{a \in X} \mathrm{Res}(\omega_a, a).$$

Both definitions of the residue are related:

5.3 Proposition. *The diagram*

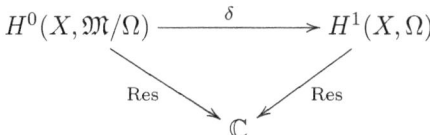

commutes.

Proof. We start with an element from $H^0(X, \mathfrak{M}/\Omega)$. As we have seen there exists an open covering (U_i) and meromorphic differentials ω_i on U_i such that $\omega_{ij} := \omega_i - \omega_j$ is holomorphic on $U_i \cap U_j$. The family ω_{ij} is a Čech one-cocycle with coefficients in Ω and defines an element of $H^1(X, \Omega)$. This is the image of $(\omega_i) \in H^0(X, \mathfrak{M}/\Omega)$ under the combining map $\delta : H^0(X, \mathfrak{M}/\Omega) \longrightarrow H^1(X, \Omega)$. We can consider the forms (ω_{ij}) also as a Čech-one-cocycle of $\mathcal{A}^{1,0}$. Since the first cohomology of this sheaf vanishes, there exist elements $\eta_i \in \mathcal{A}^{1,0}(U_i)$ such that

$$\eta_i - \eta_j = \omega_{ij} \quad (= \omega_i - \omega_j)$$

The advantage of the η_i compared to the ω_i is that they have no poles in U_i. The prize which we have to pay is that they are not holomorphic. Since $\omega_i - \omega_j$ is holomorphic, we have $d\omega_i = d\omega_j$ and hence $d\eta_i = d\eta_j$ on $U_i \cap U_j$. Hence the $d\eta_i$ glue to a differential $\gamma \in \mathcal{A}^2(X)$. This defines a cohomology class in

$H^1(X,\Omega) = \mathcal{A}^2(X)/\partial\mathcal{A}^{1,0}$. One can check (we leave this to the reader) that this cohomology class corresponds to the one-cocycle (ω_{ij}).

We have to compute $\int_X \gamma$ and to relate it to the residues of (ω_i). Hence we consider the finite set of poles S. The differences $\eta_i - \omega_i$ are smooth on $X' = X - S$ and glue to a differential $\eta \in \mathcal{A}^{1,0}(X')$. We have $d\eta = -\gamma$ on X'. Hence $d\eta$ (but not η) has a smooth extension to X.

We now consider the integral $\int_X \gamma$. We cut out the disk of radius ε around each $a \in S$ (with respect to some chart). We take ε small enough such that the disks are pairwise disjoint. We denote the complement of the disks by $X(\varepsilon)$. We have

$$\int_X \gamma = \lim_{\varepsilon \to 0} \int_{X(\varepsilon)} \gamma.$$

Now Stokes' formula applies and gives

$$\int_{X(\varepsilon)} \gamma = \sum_{a \in S} \oint_\alpha \eta,$$

where α denotes a small circle around a (mathematically negative orientation).

The proof of Proposition 5.3 will be complete if we show

$$\lim_{\varepsilon \to 0} \oint_\alpha \eta = 2\pi i \operatorname{Res} \omega_i \qquad (a \in U_i).$$

Here we have chosen i such that $a \in U_i$. On a small punctured neighborhood of a we have $\eta = \eta_i - \omega_i$. Since the integral of $-\omega_i$ produces the residue, it remains to show

$$\lim_{\varepsilon \to 0} \oint_\alpha \eta_i = 0.$$

But this is clear since η_i is smooth on U_i (including a). □

5.4 Proposition. *The residue map $H^1(X,\Omega) \to \mathbb{C}$ is different from zero.*

Proof. Because of Proposition 5.3 it is sufficient to show that the map $H^0(X,\mathfrak{M}/\Omega) \to \mathbb{C}$ is not zero. For this one takes a point $a \in X$ and a meromorphic differential ω_1 on a small open neighborhood U_1 which has a simple pole at a and no other pole. Then we consider the zero form ω_2 on $U_2 = X - \{a\}$. Both together glue to a global section of $\mathfrak{M}/\Omega$. Its residue is different from 0. □

6. Serre duality

Let $\mathcal{M}$ be a vector bundle on the compact Riemann surface X. We consider the vector bundle $\mathcal{H}om(\mathcal{M}, \Omega)$. There is a natural $\mathcal{O}_X$-bilinear map

$$\mathcal{M} \times \mathcal{H}om(\mathcal{M}, \Omega) \longrightarrow \Omega.$$

Locally it is given by $(m, l) \mapsto l(m)$. For a fixed global section from $\mathcal{H}om(\mathcal{M}, \Omega)$ this induces a map

$$\mathcal{M} \longrightarrow \Omega.$$

Taking cohomology we get a map

$$H^1(X, \mathcal{M}) \longrightarrow H^1(X, \Omega).$$

Varying the global section we get a bilinear map

$$H^0(X, \mathcal{H}om(\mathcal{M}, \Omega)) \times H^1(X, \mathcal{M}) \longrightarrow H^1(X, \Omega).$$

Combining it with the residue map we get

$$H^0(X, \mathcal{H}om(\mathcal{M}, \Omega)) \times H^1(X, \mathcal{M}) \longrightarrow \mathbb{C}.$$

This can be considered as a linear map

$$\boxed{H^0(X, \mathcal{H}om(\mathcal{M}, \Omega)) \longrightarrow H^1(X, \mathcal{M})^*}$$

where V^* denotes the dual of a vector space ($V^* = \mathrm{Hom}_{\mathbb{C}}(V, \mathbb{C})$). We call this map the *duality map*.

6.1 Serre duality. *For a vector bundle $\mathcal{M}$ on a compact Riemann surface the duality map*

$$H^0(X, \mathcal{H}om(\mathcal{M}, \Omega)) \longrightarrow H^1(X, \mathcal{M})^*$$

is an isomorphism.

Corollary. $\dim H^1(X, \mathcal{M}) = \dim H^0(X, \mathcal{H}om(\mathcal{M}, \Omega))$.

Proof. In a first step we prove the injectivity of the duality map. Let $f \in H^0(X, \mathcal{H}om(\mathcal{M}, \Omega))$ be a non-zero element. This is a non-zero $\mathcal{O}$-linear map $f : \mathcal{M} \longrightarrow \Omega$. We have to show that its image in $H^1(X, \mathcal{M})^*$ is not zero. For this, we use the sequence

$$0 \longrightarrow \mathrm{Kernel}(f) \longrightarrow \mathcal{M} \longrightarrow f(\mathcal{M}) \longrightarrow 0.$$

We know that $f(\mathcal{M})$ is a line bundle (Lemma 3.4) and Kernel(f) is a vector bundle (Coherence Lemma 3.5). Since $H^2(X, \text{Kernel}(f)) = 0$ we get that $H^1(X, \mathcal{M}) \to H^1(X, f(\mathcal{M}))$ is surjective.

Now we use the sequence
$$0 \longrightarrow f(\mathcal{M}) \longrightarrow \Omega \longrightarrow \Omega/f(\mathcal{M}) \longrightarrow 0.$$
Since $\Omega/f(\mathcal{M})$ is a skyscraper sheaf, its first cohomology group vanishes and we get the surjectivity of $H^1(X, f(\mathcal{M})) \to H^1(X, \Omega)$. As a consequence, the composition $H^1(X, \mathcal{M}) \longrightarrow H^1(X, \Omega)$ is surjective. Because of Proposition 5.4 the composition with the residue map $H^1(X, \Omega) \to \mathbb{C}$ is non-zero. It is easy to check that this is the image of f under the duality map. So we have shown that the image of f under the duality map is not 0. This proves the injectivity of the duality map. □

Next we will proof the surjectivity of the duality map. This is more involved, since it needs some control that $H^1(X, \mathcal{M})$ is not to big. The idea is to consider twists $\mathcal{M}(n)$. Recall that to define $\mathcal{M}(n)$, one has to choose a point a and then to consider the divisor $n(a)$ which is concentrated on a with multiplicity n.

6.2 Lemma. *For $n \to \infty$ the asymptotic formula*
$$\dim H^0(X, \mathcal{O}(n)) = n + O(1)$$
holds.

Proof. From Lemma 3.2 we know that the dimension of $H^1(X, \mathcal{O}(n))$ is bounded. Now the claim follows from the Riemann–Roch formula. □

The line bundle $\mathcal{O}(-n)$ is contained in $\mathcal{O}$ when $n \geq 0$. Hence $\mathcal{M}(-n)$ can be considered as subsheaf of $\mathcal{M}$ for $n \geq 0$. In particular,
$$\dim H^0(X, \mathcal{M}(-n)) \leq \dim H^0(X, \mathcal{M}).$$
We use the formula $\deg(\mathcal{M}(-n)) = \deg(\mathcal{M}) - n \cdot \text{rank}\mathcal{M}$ (Lemma 4.5). Of course $\text{rank}(\mathcal{M}) = \text{rank}(\mathcal{M}(n))$. Hence Riemann–Roch gives
$$\dim H^1(X, \mathcal{M}(-n)) =$$
$$\dim H^0(X, \mathcal{M}(-n)) + n \cdot \text{rank}\mathcal{M} - \deg(\mathcal{M}) - \text{rank}(\mathcal{M})(1-g).$$
As we have seen, $\dim H^0(X, \mathcal{M}(-n))$ is bounded for $n > 0$.

6.3 Lemma. *For $n \to \infty$ the asymptotic formula*
$$\dim H^1(X, \mathcal{M}(-n)) = n \cdot \text{rank}\mathcal{M} + O(1)$$
holds.

Similarly we get the asymptotic behavior of $\dim H^0(X, \mathcal{Hom}(\mathcal{M}(-n), \Omega))$. By Lemma 4.5 one has $\deg \mathcal{Hom}(\mathcal{M}(-n), \Omega) = n \cdot \text{rank}(\mathcal{M}) - \deg \mathcal{M} + \deg(\Omega)$. The inclusion $\mathcal{M}(-n) \hookrightarrow \mathcal{M}$ gives a surjection $\mathcal{Hom}(\mathcal{M}, \Omega) \to \mathcal{Hom}(\mathcal{M}(-n), \Omega)$ and this induces a surjection in the first cohomology (vanishing of H^2). We get the boundedness of $\dim H^1(X, \mathcal{Hom}(\mathcal{M}(-n), \Omega))$. Now Riemann–Roch shows the following asymptotic formula which, together with Lemma 6.3, can be considered as a weak form of the duality theorem.

§6. Serre duality

6.4 Lemma. *For $n \to \infty$ the asymptotic formula*

$$\dim H^0(X, \quad (\mathcal{M}(-n), \Omega)) = n \cdot \mathrm{rank}\mathcal{M} + O(1)$$

holds.

Next we have to investigate the behavior of the duality map under twisting. From the exact sequence $0 \to \mathcal{M}(-1) \to \mathcal{M} \to \mathcal{M}/\mathcal{M}(-1) \to 0$ we obtain the exact sequence

$$H^0(X, \mathcal{M}/\mathcal{M}(-1)) \longrightarrow H^1(X, \mathcal{M}(-1)) \longrightarrow H^1(X, \mathcal{M}).$$

Dualizing gives the exact sequence

$$\boxed{H^1(X, \mathcal{M})^* \longrightarrow H^1(X, \mathcal{M}(-1))^* \longrightarrow (\mathcal{M}_a/\mathcal{M}(-1)_a)^*}$$

We have to construct a second sequence: the imbedding $\mathcal{M}(-1) \hookrightarrow \mathcal{M}$ gives a map

$$o(\mathcal{M}, \Omega) \longrightarrow o(\mathcal{M}(-1), \Omega).$$

Now we use the fact that an $\mathcal{O}$-linear map of $\mathcal{O}$-modules $\mathcal{M}_1 \to \mathcal{M}_2$ induces a natural map $\mathcal{M}_1 \otimes_{\mathcal{O}} \mathcal{L} \to \mathcal{M}_2 \otimes_{\mathcal{O}} \mathcal{L}$. We can read this as an isomorphism $(\mathcal{M}_1, \mathcal{M}_2) \cong (\mathcal{M}_1 \otimes_{\mathcal{O}} \mathcal{L}, \mathcal{M}_2 \otimes_{\mathcal{O}} \mathcal{L})$. In particular, we can identify

$$o(\mathcal{M}(-1), \Omega) = \quad o(\mathcal{M}, \Omega(1)).$$

Let $\mathcal{M}'$ be the sheaf of meromorphic differentials which are holomorphic outside a and have possibly in a a pole of order at most one. There is an obvious isomorphism $\Omega(1) \to \mathcal{M}'$ which sends $\omega \otimes f$ to $f\omega$. The residue map $\mathcal{M}'_a \to \mathbb{C}$ induces a map $\Omega(1)_a \to \mathbb{C}$.

Let $\mathcal{M}(-1) \to \Omega$ be an $\mathcal{O}$-linear map. As we mentioned, this corresponds to a map $\mathcal{M} \to \Omega(1)$. It induces a map $\mathcal{M}_a \to \Omega(1)_a$. We can combine it with the residue map above to produce a map $\varphi : \mathcal{M}_a \to \mathbb{C}$. This map is zero on $\mathcal{M}_a(-1)$. Hence it induces a linear map $\mathcal{M}_a/\mathcal{M}(-1)_a \to \mathbb{C}$. We can read this construction as a map

$$H^0(X, \quad o(\mathcal{M}(-1), \Omega) \longrightarrow (\mathcal{M}_a/\mathcal{M}(-1)_a)^*.$$

By means of the natural embedding $\mathcal{M}(-1) \to \mathcal{M}$ we obtain the sequence

$$\boxed{H^0(X, \quad o(\mathcal{M}, \Omega) \longrightarrow H^0(X, \quad o(\mathcal{M}(-1), \Omega) \longrightarrow (\mathcal{M}_a/\mathcal{M}(-1)_a)^*}$$

A local consideration shows that it is exact. The two framed sequences can be combined to a commutative diagram.

6.5 Lemma. *The diagram*

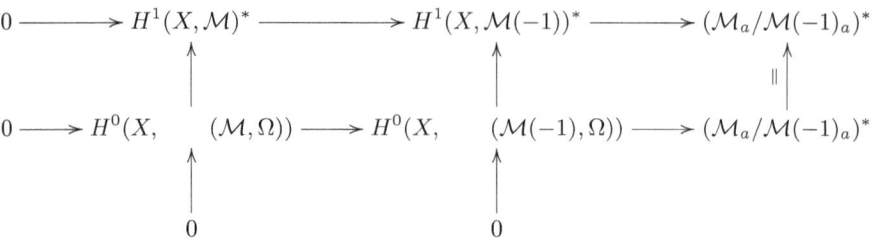

is commutative. Its lines and columns are exact.

Proof. The proof of the commutativity is leaft to the reader. It is clear that the rows are exact. The exactness of the columns follows from the injectivity of the duality map. □

Let $\lambda \in H^1(X, \mathcal{M})^*$. Our goal is to show that it is contained in the image of $H^0(X, \mathcal{H}om_\mathcal{O}(\mathcal{M}, \Omega))$. We make the weaker assumption that the image of λ in $H^1(X, \mathcal{M}(-1))^*$ comes from $H^0(\mathcal{H}om_\mathcal{O}(\mathcal{M}(-1), \Omega))$. Simple diagram chasing in Lemma 6.5 shows that then λ itself comes from $H^0(X, \mathcal{H}om_\mathcal{O}(\mathcal{M}, \Omega))$. We can apply this several times and obtain the following result.

6.6 Lemma. *Let $\lambda \in H^1(X, \mathcal{M})^*$ and $D \geq 0$ a divisor. We write $\mathcal{M}_{-D} := \mathcal{M} \otimes \mathcal{O}_{-D}$. (This is imbedded in $\mathcal{M}$.) Assume that the image of λ under $H^1(X, \mathcal{M})^* \hookrightarrow H^1(X, \mathcal{M}_{-D})^*$ comes from $H^0(X, \mathcal{H}om_\mathcal{O}(\mathcal{M}_{-D}, \Omega))$. Then λ comes from $H^0(X, \mathcal{H}om_\mathcal{O}(\mathcal{M}, \Omega))$.*

We need a last preparation of the proof of the duality theorem. Assume that a function $f \in H^0(X, \mathcal{O}(n))$ is given, i.e. a meromorphic function which is holomorphic outside a and which has in a at most a pole of order $n > 0$. Multiplication by f defines a $\mathcal{O}$-linear map $f : \mathcal{M}(-n) \to \mathcal{M}$. (We considered up to now this map for $f = 1$.) This map induces a linear map

$$f : H^1(X, \mathcal{M})^* \longrightarrow H^1(X, \mathcal{M}(-n))^*.$$

Now we fix a non-zero element $\lambda \in H^1(X, \mathcal{M})^*$. (Recall that our goal is to prove that it is in the image of the duality map.) We use it to define the map

$$H^0(X, \mathcal{O}(n)) \longrightarrow H^1(X, \mathcal{M}(-n))^*, \quad f \longmapsto f(\lambda).$$

6.7 Lemma. *For non zero $\lambda \in H^1(X, \mathcal{M}(-m))^*$, the map*

$$H^0(X, \mathcal{O}(n)) \longrightarrow H^1(X, \mathcal{M}(-n))^*, \quad f \longmapsto f(\lambda),$$

is injective.

Proof. We have to show that $f(\lambda) \neq 0$ for $f \neq 0$. This can also be interpreted as follows. For fixed $f \neq 0$ the map $f : H^1(X, \mathcal{M})^* \longrightarrow H^1(X, \mathcal{M}(-n))^*$ is injective, or equivalently, the map $f : H^1(X, \mathcal{M}(-n)) \longrightarrow H^1(X, \mathcal{M})$

is surjective. This follows from the long exact cohomology sequence since $\mathcal{M}/f(\mathcal{M}(-n))$ is a skyscraper sheaf. □

Now we consider the triangle

$$H^0(X, \mathcal{O}(n)) \xrightarrow{f \mapsto f(\lambda)} H^1(X, \mathcal{M}(-n))^*$$
$$\uparrow$$
$$H^0(\operatorname{Hom}(\mathcal{M}(-n)), \Omega).$$

We know that both arrows are injective. We use the asymptotic formulas in the Lemmas 6.3, 6.4 and 6.2. They show that the images of $H^0(X, \mathcal{O}(n))$ and $H^0(\operatorname{Hom}(\mathcal{M}(-n)), \Omega)$ for large n have a non-zero intersection in the space $H^1(X, \mathcal{M}(-n))$. Therefore, for large n, there must exist a non-zero $f \in H^0(X, \mathcal{O}(n))$ such that $f(\lambda)$ is in the image of the vertical arrow (=duality map for $\mathcal{M}(-n)$). We consider the divisor $D = n(a) + (f)$ with the property $D \geq 0$. Hence $\mathcal{M}_{-D}$ is naturally imbedded in $\mathcal{M}$. Multiplication by f gives already a map $f : \mathcal{M}_{-D} \to \mathcal{M}$. Hence the map $f : H^1(X, \mathcal{M})^* \to H^1(X, \mathcal{M}(-n))^*$ factors as

$$H^1(X, \mathcal{M})^* \xrightarrow{\text{natural imbedding}} H^1(X, \mathcal{M}_{-D})^* \xrightarrow{f} H^1(X, \mathcal{M}(-n))^*$$
$$\lambda \mapsto \lambda \mapsto f(\lambda)$$

In the same way we get a factorization

$$H^0(X, \operatorname{Hom}(\mathcal{M}, \Omega)) \xrightarrow{\text{nat. i.}} H^0(X, \operatorname{Hom}(\mathcal{M}_{-D}, \Omega)) \xrightarrow{f} H^0(X, \operatorname{Hom}(\mathcal{M}(-n), \Omega)).$$

This shows that $\lambda \in \operatorname{Hom}(\mathcal{M}_{-D}, \Omega)$ is contained in the image of the duality map. Now Lemma 6.6 applies and shows that already $\lambda \in \operatorname{Hom}(\mathcal{M}, \Omega)$ is contained in the duality map. This completes the proof of the duality theorem. □

7. Some comments on the Riemann–Roch theorem

The Riemann–Roch theorem can now be written in the cohomology free form

$$\dim H^0(X, \mathcal{M}) - \dim H^0(X, \operatorname{Hom}(\mathcal{M}, \Omega)) = \deg(\mathcal{M}) + \operatorname{rank}(\mathcal{M})(1 - g).$$

We formulate it in its classical form for divisors. A divisor K is called *canonical* if the associated line bundle is isomorphic to Ω. One can get a canonical divisor as follows: Let ω be a meromorphic 1-form. It is clear how to associate to ω a divisor $D = (\omega)$ which describes the poles and zeros of ω. This is obviously a canonical divisor, since the sections of $\mathcal{O}_D$ are the meromorphic functions f with the property that $f\omega$ is holomorphic. This gives an isomorphism $\mathcal{O}_D \cong \Omega$. By the way, a non vanishing meromorphic 1-form exists. One can take one of the form df, where f is a non-constant meromorphic function, which exists by Theorem 3.7.

Using the notation

$$l(D) := \dim H^0(X, \mathcal{O}_D) = \dim\{\, f \text{ meromorphic on } X;\quad (f) \geq -D\,\},$$

we now get

7.1 Classical Riemann–Roch. *Let D be a divisor on a compact Riemann surface and let be K a canonical divisor. Then*

$$l(D) - l(K - D) = \deg(D) + 1 - g.$$

We consider some special cases: let D be the zero divisor. Then $l(D) = 1$, $\deg(D) = 0$ and we obtain a different definition for the genus.

7.2 Theorem. *One has $\dim \Omega(X) = g$.*

We now take $D = K$ in the Riemann–Roch formula. We get $g - 1 = \deg(D) + 1 - g$:

7.3 Theorem. *The degree of a canonical divisor (or of Ω) is $2g - 2$.*

If f is a non-zero meromorphic function with the property $(f) + D \geq 0$, then $\deg(D) \geq 0$. Hence $l(D) = 0$ for $\deg(D) < 0$. Applying this to $K - D$ we get:

7.4 Theorem. *Let D be a divisor with $\deg(D) > 2g - 2$. Then*

$$l(D) = \deg(D) + 1 - g.$$

Finally, we derive a formula which allows to compute the genus. Assume that $f : X \to Y$ is a non-constant holomorphic map of compact Riemann surfaces. Let $a \in X$. Recall that one can define the multiplicity $\mathrm{Ord}(f, a)$ of f at a. We call

$$b(f, a) := \mathrm{Ord}(f, a) - 1$$

the *ramification order*. This is zero if and only if f is locally biholmorphic at a. We denote by

$$b(f) = \sum_{a \in X} b(f, a)$$

the *total ramification order*. The sum is of course finite.

§7. Some comments on the Riemann–Roch theorem

7.5 Riemann–Hurwitz ramification formula. Let $f : X \to Y$ be a non-constant holomorphic map of compact Riemann surfaces. Let n be the covering degree of f (see Proposition V.5.8). Then the following relation between the genus g_X of X and g_Y of Y holds:

$$g_X = \frac{b(f)}{2} + n(g_Y - 1) + 1.$$

We just give a short hint for the proof. Consider a meromorphic function g on Y and compare the divisors of the differentials dg on Y and $d(g \circ f)$ on X. Use Theorem 7.3. □

We treat some simple examples.

1) Since $H^1(\bar{\mathbb{C}}, \mathcal{O}) = 0$ (Theorem IV.4.7), we have that the genus of the Riemann sphere is zero.

2) A meromorphic differential on $\bar{\mathbb{C}}$ is given by dz. It has pole of order 2 at ∞ (use the chart $1/z$ and the fact that the derivative of $1/z$ is $-1/z^2$) and no further poles or zeros. Hence the degree of the canonical divisor is -2 in concordance with the expression $2g - 2$.

3) Consider the map $\bar{\mathbb{C}} \to \bar{\mathbb{C}}$, $z \mapsto z^2$. There are two ramification points $0, \infty$ of ramification order 1. The covering degree is 2, the total ramification number is also 2. The Riemann–Hurwitz formula gives $0 = 1 + 2(0-1) + 1$ which is correct.

4) Consider a complex torus $X = \mathbb{C}/L$. The differential dz on $\mathbb{C}$ is invariant under translations. Hence it gives a holomorphic differential on X. It has no poles and zeros, hence its degree is zero. From $2g - 2 = 0$ we get $g = 1$. Hence the genus of a torus is one.

5) Another way to see this is to use the Weierstrass $\wp$-function. This is a holomorphic map $\mathbb{C}/L \to \bar{\mathbb{C}}$ with 4 ramification points (the zeros of $\wp'$ and ∞). The covering degree is 2, the total ramification number is 4. The Riemann–Hurwitz formula gives $g = 2 + 2(-1) + 1 = 1$.

6) At the end of Chapt. V and Sect. 6, we explained the Riemann surface of the "function $\sqrt{P(z)}$", where P is a polynomial without multiple zeros. Let the degree of P be $n = 2g+1$ or $n = 2g+2$. We explained that this Riemann surface is a two fold covering $f : X \to \bar{\mathbb{C}}$. The number b of ramification points (all of order two) turned out to be $n \leq b \leq n+1$. Since we know from the Riemann–Hurwitz formula 7.5 that the total ramification order is even, we can conclude that in both cases $b = 2g$ is the correct number. From the Riemann–Hurwitz formula we get:

The genus of the Riemann surface "$\sqrt{P(z)}$" is g, where the degree of P is $2g+1$ or $2g+2$.

We consider the projection $z : X \to \bar{\mathbb{C}}$. Then dz is a meromorphic differential on X. Recall that $\sqrt{P(z)}$ has been defined as a meromorphic

function on X. Hence we can consider the meromorphic differentials on X
$$\frac{z^k dz}{\sqrt{P(z)}}.$$
If one goes carefully through the construction of X, one can work out when this differential has no poles. The result is that this is the case if and only if
$$0 \leq k < g.$$
Since these differentials are linearly independent, they must give a basis of $\Omega(X)$ by Theorem 7.2.

The integration of these integrals is a major problem. In the case $g = 0$ one has so-called circle integrals. The integration is elementary and leads to the functions sin and cos. The case $g = 1$ is more involved. The integrals are called elliptic integrals. Their integration leads to the theory of elliptic functions. In the language of Riemann surfaces this means that one can show that X is biholomorphic equivalent to a complex torus $\mathbb{C}/L$ and that every complex torus arises in this way. (The fact that both types have genus 1 may be considered as a weak hint that this is true.) In the case $g > 1$ the integrals are called "hyperelliptic integrals" and the corresponding Riemann surfaces are called *hyperelliptic Riemann surfaces*. The integration of hyperelliptic differentials leads deep into the theory of Riemann surfaces. With their help deep open problems as the Jacobi inversion problem could be solved.

Chapter VII. The Jacobi inversion problem

1. Harmonic differentials

Recall that a function u on an open subset $U \subset \mathbb{C}$ is called *harmonic* if it satisfies
$$\Delta u := \left(\frac{\partial^2}{\partial x^2} + \frac{\partial^2}{\partial y^2}\right)u = 0.$$

A function is harmonic if and only if its real and imaginary part are harmonic. It is known from complex calculus that a real function u is harmonic if and only if it is locally the real part of an analytic function. This enables a quick generalization to Riemann surfaces.

1.1 Definition. *A real function u on a Riemann surface X is called harmonic if every point admits an open neighborhood U such that u can be written on U as real part of an analytic function $f : U \to \mathbb{C}$. An arbitrary (complex valued) function u is called harmonic if its real and imaginary part are harmonic.*

We also need the notion of a harmonic differential.

1.2 Definition. *A differential $\omega \in \mathcal{A}^1(X)$ on a Riemann surface is called harmonic if every point admits an open neighborhood U such that ω can be written on U as du with a harmonic function $u : U \to \mathbb{C}$.*

Analytic functions are harmonic. As a consequence, holomorphic differentials are also harmonic.

We define the complex conjugate $\bar{\omega}$ of a differential ω. In coordinates this is defined as
$$\overline{f\,dx + g\,dy} := \bar{f}\,dx + \bar{g}\,dy.$$

One checks $\gamma^*\bar{\omega} = \overline{\gamma^*\omega}$ for arbitrary differentiable maps. This allows to generalize complex conjugation on Riemann surfaces via charts. A differential ω is called *real* if $\omega = \bar{\omega}$. The formula $\overline{dz} = d\bar{z}$ shows that complex conjugation defines an isomorphism
$$\mathcal{A}^{1,0}(X) \xrightarrow{\sim} \mathcal{A}^{0,1}(X), \qquad \omega \longmapsto \bar{\omega}.$$

The complex conjugate of a harmonic differential is harmonic as well. Hence we see that not only holomorphic differentials are harmonic but also antiholomorphic ones. (A function or a differential is called antiholomorphic if its complex conjugate is holomorphic.)

1.3 Remark. *A harmonic differential ω can be written as the sum of a holomorphic and an antiholomorphic differential.*

Proof. First we notice that a harmonic function locally is the sum of a holomorphic and an antiholomorphic function. This shows that a harmonic differential locally can be written in the form $\omega = \omega_1 + \omega_2$ with a holomorphic differential ω_1 and an antiholomorphic differential ω_2. Since this decomposition is obviously unique, it extends to a global decomposition. □

We introduce the so-called star operator. Recall that we have the direct decomposition
$$\mathcal{A}^1(X) = \mathcal{A}^{1,0}(X) \oplus \mathcal{A}^{0,1}(X).$$

1.4 Definition. *The **star operator***
$$* : \mathcal{A}^1(X) \longrightarrow \mathcal{A}^1(X)$$
is defined by
$$*(\omega_1 + \omega_2) = \mathrm{i}(\bar{\omega}_1 - \bar{\omega}_2), \qquad \omega_1 \in \mathcal{A}^{1,0}(X), \quad \omega_2 \in \mathcal{A}^{0,1}(X).$$

Before we continue, we formulate some rules which we will use in the following:

1.5 Some rules.
a) $d(f\omega) = (df) \wedge \omega + f d\omega \qquad (f \in \mathcal{C}^\infty(X), \ \omega \in \mathcal{A}^1(X))$.
b) *The same as in* a) *but d replaced by ∂ or $\bar{\partial}$.*
c) $* * \omega = -\omega$.
d) $\overline{*\omega} = *\bar{\omega}$.
e) $d(*\omega) = \mathrm{i}\partial\bar{\omega}$ for $\omega \in \mathcal{A}^{1,0}(X)$, $\quad d(*\omega) = -\mathrm{i}\bar{\partial}\omega$ for $\omega \in \mathcal{A}^{0,1}(X)$.
f) $*\partial f = \mathrm{i}\bar{\partial}\bar{f}, \quad *\bar{\partial}f = -\mathrm{i}\partial\bar{f}$.
g) $d(*df) = 2\mathrm{i}\partial\bar{\partial}\bar{f}$.

The proof can be given by direct calculations with coordinates. □

The announced new characterization of harmonic differentials uses this star operator:

1.6 Proposition. *A differential ω on a Riemann surface is harmonic if and only if*
$$d\omega = d(*\omega) = 0.$$

Proof. Let ω be harmonic. We show the two differential equations. Because of Remark 1.3 we can assume that ω is holomorphic or antiholomorphic. But then $*\omega$ is antiholomorphic or holomorphic and the statement is clear.

Now we assume conversely the two differential equations. From $d\omega = 0$ follows that ω can be locally written as df with some function f. A calculation in coordinates shows that $d(*\omega) = d(*df) = 0$ means that f is harmonic (use the rule 1.5, g)). $\square$

We denote by $\mathrm{Harm}^1(X)$ the space of all harmonic differentials on X. We reformulate Remark 1.3 in the form

$$\mathrm{Harm}^1(X) = \Omega(X) \oplus \bar{\Omega}(X).$$

Hence, for compact X, we get

$$\dim_{\mathbb{C}} \mathrm{Harm}^1(X) = 2g.$$

2. Hodge theory of compact Riemann surfaces

In the following X denotes a connected Riemann surface. For two differentials α, β on a Riemann surface X we define

$$[\alpha, \beta] = \alpha \wedge *\beta.$$

This is $\mathbb{C}$-linear in the first variable and satisfies

$$[\beta, \alpha] = \overline{[\alpha, \beta]}.$$

We express $[\alpha, \alpha]$ in local coordinates and write α for this purpose in the form

$$\alpha = f dz + g d\bar{z}.$$

Then

$$*\alpha = i(\bar{f} d\bar{z} - \bar{g} dz)$$

and hence

$$[\alpha, \alpha] = i(|f|^2 + |g|^2) dz \wedge d\bar{z} = 2(|f|^2 + |g|^2) dx \wedge dy.$$

The essential point here is that $|f|^2 + |g|^2$ is non-negative and zero only if the (smooth) functions f, g are zero.

Now we will assume that the Riemann surface X is compact. Then we can define

$$\langle \alpha, \beta \rangle = \int_X [\alpha, \beta].$$

This is a hermitian form on $\mathcal{A}^1(X)$ ($\mathbb{C}$-linear in the first variable and with the property $\langle \beta, \alpha \rangle = \overline{\langle \alpha, \beta \rangle}$). Moreover, this form is positive definite. Hence $\mathcal{A}^1(X)$ has been equipped with a structure as unitary vector space.

2.1 Proposition. *On any compact Riemann surface one has an orthogonal decomposition (with respect to $\langle \cdot, \cdot \rangle$)*

$$\mathcal{A}^{0,1}(X) = \bar{\partial} C^\infty(X) \oplus \bar{\Omega}(X).$$

Corollary. *A form σ in $\mathcal{A}^{0,1}(X)$ can be written in the form $\sigma = \bar{\partial} f$ if and only if*

$$\int_X \sigma \wedge \omega = 0$$

for all holomorphic differentials ω.

Proof. First we prove that $\bar{\partial} C^\infty(X)$ and $\bar{\Omega}(X)$ are orthogonal: we have to consider $\omega \wedge *\bar{\partial} f$ for antiholomorphic ω. Using the rules 1.5 one shows

$$\omega \wedge *\bar{\partial} f = -i d(\bar{f}\omega).$$

From Stokes formula follows $\langle \omega, df \rangle = 0$.

From the orthogonality we obtain that the natural map

$$\bar{\Omega}(X) \to \mathcal{A}^{0,1}(X)/\bar{\partial} C^\infty(X)$$

is injective. Hence for the proof of 2.1 it remains to show that both sides have the same dimension. The dimension of $\bar{\Omega}$ is g (by duality). By Dolbeault's theorem V.7.6, the right hand side is isomorphic to $H^1(X, \mathcal{O}_X)$ which by definition has also the dimension g. This completes the proof of Proposition 2.1. □

We mention that for trivial reason the spaces $\mathcal{A}^{1,0}(X)$ and $\mathcal{A}^{0,1}(X)$ are orthogonal (since $dz \wedge dz = 0$). Hence not only $\bar{\partial} C^\infty(X)$ but also $\partial C^\infty(X)$ is orthogonal to $\bar{\Omega}(X)$. We obtain that also $dC^\infty(X)$ is orthogonal to $\bar{\Omega}(X)$. Since $dC^\infty(X)$ is invariant under complex conjugation, it is also orthogonal to $\Omega(X)$. It follows that $dC^\infty(X)$ and $\mathrm{Harm}^1(X)$ are orthogonal. Let now u be a harmonic function on X. Then du is a harmonic differential which is orthogonal to $\mathrm{Harm}^1(X)$ hence to itself. This means $du = 0$ and we obtain the following proposition.

2.2 Proposition. *Every harmonic function on a compact Riemann surface is constant.*

We give another application of Proposition 2.1. Complex conjugation gives

$$\mathcal{A}^{1,0}(X) = \partial C^\infty(X) \oplus \Omega(X)$$

or

$$\mathcal{A}^1(X) = \partial C^\infty(X) \oplus \bar{\partial} C^\infty(X) \oplus \mathrm{Harm}^1(X).$$

Now we use the rule 1.5, f). It shows that

$$dC^\infty(X) + *dC^\infty(X) \subset \partial C^\infty(X) + \bar{\partial} C^\infty(X).$$

From the same rule follows also $df + *d(i\bar{f}) = \partial f$ and hence the converse inclusion. The spaces $dC^\infty(X)$ and $*dC^\infty(X)$ are also orthogonal. This follows from $df \wedge **dg = -df \wedge dg = -d(fdg)$. This gives us the following result.

§2. Hodge theory of compact Riemann surfaces

2.3 Proposition. *On a compact Riemann surface one has the orthogonal decomposition*

$$\mathcal{A}^1(X) = dC^\infty(X) \oplus *dC^\infty(X) \oplus \mathrm{Harm}^1(X).$$

We want to determine the position of $K := \mathrm{Kernel}(\mathcal{A}^1(X) \xrightarrow{d} \mathcal{A}^2(X))$ with respect to this decomposition. We claim that it is orthogonal to $*dC^\infty(X)$. For $d\omega = 0$ one gets

$$\omega \wedge *(*df) = -\omega \wedge df = d(f\omega).$$

This shows $\langle \omega, *df \rangle = 0$. On the other side $dC^\infty(X)$ and $\mathrm{Harm}^1(X)$ are contained in K. Hence K equals their sum:

2.4 Proposition. *On a compact Riemann surface one has the orthogonal decomposition*

$$\mathrm{Kernel}(\mathcal{A}^1(X) \xrightarrow{d} \mathcal{A}^2(X)) = dC^\infty(X) \oplus \mathrm{Harm}^1(X).$$

This proposition induces an isomorphism

$$\mathrm{Harm}^1(X) \cong \frac{\mathrm{Kernel}(\mathcal{A}^1(X) \xrightarrow{d} \mathcal{A}^2(X))}{dC^\infty(X)}.$$

By the theorem of de Rham the right hand side is isomorphic to $H^1(X, \mathbb{C})$. This gives us the following important theorem.

2.5 Theorem of de Rham–Hodge. *For a compact Riemann surface one has*

$$H^1(X, \mathbb{C}) \cong \mathrm{Harm}^1(X).$$

Hence the dimension of $H^1(X, \mathbb{C})$ is $2g$.

Notice that $H^1(X, \mathbb{C})$ depends only on the topological space X and not on the complex structure. Hence we see that the genus g for homeomorphic Riemann surfaces is the same. But homeomorphic Riemann surfaces need not to be biholomorphic equivalent as already the example of tori shows.

3. Integration of closed forms and homotopy

A differential ω is called *closed* if $d\omega = 0$ and *total* if it is of the form $\omega = df$. From the main theorem of calculus follows the formula

$$\int_\alpha df = \alpha(1) - \alpha(0)$$

for every smooth curve $\alpha : [0,1] \to X$. In the following it will be useful to weaken the smoothness of a curve. First of all, the curve integral $\int_\alpha \omega$ can be generalized to *piecewise* smooth curves in an obvious way. But for *closed* differentials it is possible to extend the integral to arbitrary *continuous* curves as follows. What we have to use is that closed forms are locally total by the Poincaré lemma. Hence every point admits a neighborhood U, such that $\int_\alpha \omega = 0$ for every closed curve in U when ω is closed. Let now $\alpha : [0,1] \to X$ be an arbitrary (continuous) curve. By a compactness argument we can find a partition $0 = a_0 < \cdots < a_n = 1$ and open subsets $U_1, \ldots, U_n$, biholomorphically equivalent to disks, such that

$$\alpha([a_{i-1}, a_i]) \subset U_i \qquad (1 \leq i \leq n)$$

and such that ω is total in U_i. Then we combine $\alpha(a_{i-1})$ and $\alpha(a_i)$ by some smooth curve inside U_i. This defines a new piecewise smooth curve β with the same origin and end as α. It is easy to see that $\int_\beta \omega$ is independent of the choice of β. (Here we use that ω is total in U_i.) Hence we can define

$$\int_\alpha \omega := \int_\beta \omega.$$

Let $Q \subset \mathbb{C}$ be a compact rectangle, parallel to the axes, and let $H : Q \to X$ be a continuous map. In an obvious way there can be defined a closed curve α which runs through the image of the boundary of Q. We claim

$$\int_\alpha \omega = 0 \qquad (\omega \text{ closed}).$$

The proof is very simple. One divides Q into small rectangles such that the image of each of them is contained in an open subset on which ω is total. Then the corresponding integral for the small rectangles are zero. Summing up all the integrals over the small rectangles, one gets obviously the integral over the original rectangle. We formulate a special case of this:

3.1 Definition. *Two curves* $\alpha : [0,1] \to X$ *and* $\beta : [0,1] \to X$ *with the property* $\alpha(0) = \beta(0)$ *and* $\alpha(1) = \beta(1)$ *are called* **homotopic** *if there exists a continuous map* $H : [0,1] \times [0,1] \to X$ *such that*

$$H(t, 0) = \alpha(t), \quad H(t, 1) = \beta(t)$$

and

$$H(0, s) = \alpha(0), \quad H(1, s) = \alpha(1).$$

The above observation shows:

§4. Periods

3.2 Proposition. *Let α, β be two homotopic curves. Then*

$$\int_\alpha \omega = \int_\beta \omega$$

for all closed differentials ω.

In the special case, where ω is holomorphic, this is called the homotopy version of the Cauchy integral theorem.

The fundamental group

We fix a point $a \in X$ and consider now closed curves which start from a and end in a. Homotopy defines an equivalence relation on this set. The set of all homotopy classes is denoted by $\pi_1(X, a)$. It is easy to show that composition of curves defines a well-defined product in $\pi_1(X, a)$. It is not difficult to show that this product makes $\pi_1(X, a)$ to a group. Details can be found in [Fr1], Chap. III, Appendix A. This group is called the fundamental group of (X, a).

3.3 Remark. *Let ω be a closed differential on X. Then integration defines a homomorphism*

$$\pi_1(X, a) \longrightarrow \mathbb{C}, \quad \alpha \longmapsto \int_\alpha \omega.$$

4. Periods

Let ω be a (smooth) closed differential on a Riemann surface. A complex number C is called a *period* of ω if there exists a closed curve α with the property

$$C = \int_\alpha \omega.$$

To explain the notion "period" we consider the case of a torus $X = \mathbb{C}/L$ and "$\omega = dz$". Every closed curve α in X lifts to a curve $\beta : [0, 1] \to \mathbb{C}$. Then

$$\int_\alpha dz = \beta(1) - \beta(0).$$

Since α is closed we have $\beta(0) \equiv \beta(1) \mod L$. Hence we have that the periods of dz are precisely the elements of L.

It is sometimes necessary to choose a base point $a \in X$ and to consider only curves which start from a. If α is an arbitrary closed curve, one can combine a with $\alpha(0)$, then run through α and then go back the same way as one started. This shows that each period can be obtained from a curve starting and ending in a.

The structure of the fundamental group of a compact surface can be determined by topological methods. A result, which can be obtained without any further theory, is the following lemma.

4.1 Lemma. *Let $S \subset X$ be a finite subset of a compact Riemann surface, $a \in X - S$. The fundamental group $\pi_1(X - S, a)$ is countable.*

Proof. If $X = \bar{\mathbb{C}}$, this can be seen as follows. Take the base point a to be rational. This means that real and imaginary part are rational numbers. It is easy to see that each closed curve in $\bar{\mathbb{C}} - S$ with origin a is homotopic to a piecewise linear curve with rational vertices. This is a countable set of curves.

The general case can be settled similarly by choosing a non-constant meromorphic function $f : X \to \bar{\mathbb{C}}$. We call a point in X rational if its image in $\bar{\mathbb{C}}$ is rational and we call a curve in X rational if the composition with f is piecewise linear with rational vertices. It is not difficult to prove that the fundamental group of $X - S$ can be generated by rational curves. We don't give the details. We just mention that a convenient way is to use the curve- and homotopy lifting for $X - f^{-1}A \to \mathbb{C} - A$ where A is a finite subset of $\mathbb{C}$. The point is that this map is a covering in the strong sense of topology (see for example [F1], Lemma I.3.13) and that such coverings have the curve and homotopy lifting property ([Fr1], Proposition III.5.6). □

The importance of the periods shows the following remark.

4.2 Remark. *A closed differential ($d\omega = 0$) is total, i.e. of the form $\omega = df$ if and only if all its periods vanish.*

Proof. The main theorem of calculus says

$$\int_\alpha df = f(\alpha(1)) - f(\alpha(0)).$$

This shows that the periods of df vanish. To prove the converse we *define*

$$f(x) := \int_a^x \omega.$$

Here a is a fixed chosen base point. The integral is understood as a curve integral along a curve which connects a with x. Since the periods vanish, this integral is independent of the choice of this curve. A local computation shows that $df = \omega$. □

Harmonic differentials are closely tied to their periods:

§4. Periods

4.3 Proposition. *A harmonic differential on a compact Riemann surface vanishes if all its periods are zero.*

Proof. Assume that ω is a harmonic differential whose periods vanish. Then $\omega = df$ by Remark 4.2. Now Proposition 2.3 shows that ω is orthogonal to $\mathrm{Harm}^1(X)$ hence to itself. This shows $\omega = 0$. □

A variant of Proposition 4.3 states:

4.4 Proposition. *A holomorphic differential on a compact Riemann surface vanishes if the real parts of all its periods are zero.*

Proof. From Proposition 4.3 follows that $\mathrm{Re}\,\omega = (\omega + \bar\omega)/2$ is zero. Locally ω can be written as df with a holomorphic function. It follows that f has constant real part. But then f is constant and $\omega = 0$. □

We now want to consider the periods of *all* holomorphic differentials together. This leads to the following definition.

4.5 Definition. *A $\mathbb{C}$-linear form $l : \Omega(X) \to \mathbb{C}$ is called a **period** of the compact Riemann surface X if there exists a closed curve α with the property*

$$l(\omega) = \int_\alpha \omega.$$

The set L of all periods is a subset of the dual space $\Omega(X)^*$. Actually it is an additive subgroup $L \subset \Omega(X)^*$. Hence we can consider the factor group

$$\mathrm{Jac}(X) := \Omega(X)^*/L.$$

We will show later that L is a lattice and therefore $\mathrm{Jac}(X)$ is a torus of real dimension $2g$. This will be explained in detail. Here we take it just as justification to call $\mathrm{Jac}(X)$ the *Jacobian variety*. We mention that from Lemma 4.1 follows that L is a countable set.

4.6 Definition of the Abel–Jacobi map. *Let X be a Riemann surface with a base point a. Then the Abel–Jacobi map*

$$A : X \longrightarrow \mathrm{Jac}(X)$$

is defined as follows. For a point $x \in X$ one chooses a curve α which combines a with x. Then one considers the linear form $\omega \mapsto \int_\alpha \omega$ in $\Omega(X)^*$. The image of this linear form in $\mathrm{Jac}(X)$ is independent of the choice of α and is defined to be $A(x)$.

The Abel–Jacobi map admits certain important variants which we will all denote by the same letter A: For example, one can consider for any natural number d the d-fold cartesian product $X^d := X \times \ldots \times X$ and define

$$A : X^d \longrightarrow \mathrm{Jac}(X), \qquad A(\xi_1, \ldots, \xi_d) = A(\xi_1) + \cdots + A(\xi_d).$$

Here we used that $\mathrm{Jac}(X)$ carries a natural structure as abelian group.

A closely related extension is obtained as follows. Consider a divisor D on X. The we define

$$A(D) := \sum_{a \in X} D(x) A(x).$$

This is obviously a homomorphism

$$A : \mathrm{Div}(X) \longrightarrow \mathrm{Jac}(X).$$

Of course this homomorphism depends on the choice of the base point a. But now we restrict A to the subgroup $\mathrm{Div}^0(X)$ of divisors if degree 0. Then we get:

4.7 Remark. *The Abel–Jacobi map* $A : \mathrm{Div}^0(X) \longrightarrow \mathrm{Jac}(X)$ *restricted to the divisors of degree zero is independent of the choice of the base point.*

Proof. We take another base point and choose a fixed curve β which combines a and b. We use this curve to transform closed curves with origin a to curves with origin b. Then the two Abel–Jacobi map differ by

$$\sum_{x \in X} D(x) \int_\beta \omega.$$

This is zero since D has degree zero. □

The Abel–Jacobi map on $\mathrm{Div}^0(X)$ can be defined in a slightly different way:

4.8 Remark. *Let* $D = (a_1) + \cdots + (a_n) - (b_1) - \cdots - (b_n)$ *be a divisor (of degree zero) on the compact Riemann surface X. Let γ_i be curves which join a_i with b_i. Then $A(D)$ is represented by the linear form*

$$\omega \longmapsto \sum_{i=1}^g \int_{\gamma_i} \omega.$$

Proof. One chooses a base point and joins a and a_i by curves α_i. Then one defines curves from a to b_i by joining α_i and γ_i. □

Special divisors of degree 0 are the principal divisors (f). Recall that they define a subgroup $\mathcal{H}(X) \subset \mathrm{Div}^0(X)$. The factor group

$$\mathcal{D}^0(X)/\mathcal{H}(X)$$

can be identified with a subgroup $\mathrm{Pic}^0(X)$ of the Picard group (Proposition VI.3.1). It can be considered as the group of all isomorphy classes of line bundles of degree 0.

§4. Periods 107

4.9 Proposition. *The image of a principal divisor in $\mathrm{Jac}(X)$ under the Abel–Jacobi map is zero. Hence it induces a homomorphism*

$$A : \mathrm{Pic}^0(X) \longrightarrow \mathrm{Jac}(X).$$

Proof. Let f be a non-constant meromorphic function on the compact Riemann surface X. We consider it as a map $f : X \to \bar{\mathbb{C}}$. For each $z \in \bar{\mathbb{C}}$ we consider the fibre $D_z := f^{-1}(z)$. Recall that we can talk about the multiplicity of an f at a point from X. Hence D_z can be considered naturally as a divisor. We will show that all D_z have the same image in $\mathrm{Jac}(X)$. (This proves 4.9, since $(f) = D_0 - D_\infty$.) We have to show that the map

$$\bar{\mathbb{C}} \longrightarrow \mathrm{Jac}(X), \quad z \longmapsto A(D_z),$$

is constant. What we will actually prove is, that this map is *locally liftable*.
A map $h : \bar{\mathbb{C}} \longrightarrow \mathrm{Jac}(X)$ is called locally liftable if every point of $\bar{\mathbb{C}}$ admits an open neighborhood U such that $h|U$ lifts to a holomorphic map $H : U \to \Omega(X)^*$.
It should be clear what "holomorphic map" here means. For example, one can say that $H(z)(\omega)$ is holomorphic in the usual sense for every $\omega \in \Omega(X)$. Before we proof the lifting property, we show that it will solve our problem:

4.10 Lemma. *Every locally liftable map $h : \bar{\mathbb{C}} \to \mathrm{Jac}(X)$ is constant.*

Proof. We fix a form $\omega \in \Omega(X)$. For a local lift H defined on some U we consider the holomorphic differential $\omega_H = d(H(z)(\omega))$ on U. Let G be a local lift on some other V. Then we have on the intersection

$$H(z) = G(z) + \lambda(z), \quad \lambda(z) \in L.$$

The function $\lambda(z)$ is a holomorphic function whose values lie in the countable set L (use Lemma 4.1). This implies that λ is locally constant. So ω_H and ω_G coincide on the intersection. Hence they glue to a holomorphic differential on $\bar{\mathbb{C}}$. Since every holomorphic differential on the sphere is zero, we get that h is locally constant. Since $\bar{\mathbb{C}}$ is connected, we get that h is constant. □

Proof of Proposition 4.9 continued. It remains to show that the map $h : \bar{\mathbb{C}} \to \mathrm{Jac} X$, $h(z) = D_z$, is locally liftable. We fix a point $b \in \bar{\mathbb{C}}$ and investigate the map h close to b. Let $b_1, \ldots, b_m$ be the points in X over b with multiplicities $k_1, \ldots, k_m$. Then $n = k_1 + \cdots + k_m$ is the covering degree of f. For the Abel–Jacobi map we need a base point $a \in X$ and curves $\alpha_1, \ldots, \alpha_n$ which combine a with the b_i. This is understood as follows. The first k_1 curves go from a to b_1, the next k_2 from a to b_2 and so on. We consider a small disk U around b. We know that then $f^{-1}(U)$ is the union of disjoint disks $U_1, \ldots, U_m$ such that $b_i \in U_i$ for $1 \leq i \leq m$. Now take a point $z \in U$ which is different from a. Since U can been taken small enough the point z will have n pairwise different inverse images, where n is the covering degree of f. The n points are distributed over the m disks $U_1, \ldots, U_m$. We can assume that the first m_1 are in U_1, the next m_2 in U_2 and so on. Now consider curves β_i, $1 \leq i \leq n$ as follows. The first m_1 ones lie in U_1 and combine b_1 with the z-s which lie in U_1. The next m_2 ones lie in U_2 and combine b_2 with the z-s which lie in U_2 and so on. Now we consider the sum of the integrals

$$\int_{\beta_i} \omega \qquad (\omega \in \Omega(X)).$$

It is independent of the choice of the β_i and depends holomorphically on z. This sum represents the difference of $A(D_b) - A(D_z)$. We can take this sum to get a local lifting of h. □

5. Abel's theorem

Abel's theorem states that the map $A : \mathrm{Pic}^0(X) \to \mathrm{Jac}(X)$ is injective. To prove this, we need some local preparation. We consider the unit disk $\mathbb{E}$.

5.1 Definition. *A C^∞-function $f : \mathbb{E} - \{0\} \to \mathbb{C} - \{0\}$ has order k at 0 if $g(z) = f(z) \cdot z^{-k}$ extends to a C^∞-function on $\mathbb{E}$ which has no zero at 0.*

We are interested in the logarithmic derivative

$$\frac{df}{f} = k\frac{dz}{z} + \frac{dg}{g}.$$

Notice that dg/g is smooth on the whole $\mathbb{E}$. Let ω be a differential on $\mathbb{E}$ with compact support. We want to consider the integral

$$\int_\mathbb{E} \frac{df}{f} \wedge \omega.$$

Since df/f is singular at 0, we have to check the existence of this integral. We write $\omega = h_1 dz + h_2 d\bar{z}$. Then

$$\frac{dz}{z} \wedge \omega = \frac{h_1}{z} dz \wedge d\bar{z} = 2\mathrm{i}\frac{h_1}{z} dx \wedge dy.$$

To compute the integral we use polar coordinates $z = re^{\mathrm{i}\varphi}$. Then

$$\int_\mathbb{E} \frac{dz}{z} \wedge \omega = 2\mathrm{i} \int_0^1 \int_0^{2\pi} \frac{h_1(re^{\mathrm{i}\varphi})}{re^{\mathrm{i}\varphi}} r dr d\varphi.$$

§5. Abel's theorem

Since the factor r cancels, there is no problem with the existence of the integral.

We assume now in addition that g has compact support. We want to integrate

$$\frac{df}{f} \wedge dg = d\left(g \cdot \frac{df}{f}\right)$$

along a circle $|z| = r$, where $r < 1$ has been chosen so close to 1 such that g vanishes for $|z| \geq r$. Since gdf/f has a singularity at the origin, we have to be careful with the application of the Stokes formula. We have to choose a small $\varepsilon > 0$ and then we can say

$$\int_{\mathbb{E}} \frac{df}{f} \wedge dg = -\oint_{|z|=\varepsilon} g \cdot \frac{df}{f}.$$

(We take the integral in the mathematical positive sense, hence we need a minus sign.) We will take the limit $\varepsilon \to 0$. Then we can replace g by the constant $g(0)$ and f by z^k. This gives:

5.2 Lemma. *Let $f : \mathbb{E} - \{0\} \to \mathbb{C} - \{0\}$ be a C^∞-function on the punctured disk which is of order k at the origin. Let g be a C^∞-function on $\mathbb{E}$ with compact support. Then*

$$\int_{\mathbb{E}} \frac{df}{f} \wedge dg = kg(0).$$

The definition of order 5.1 can be generalized to Riemann surfaces in an obvious way: if f is a differentiable function without zero in a punctured neighborhood of a point a of a Riemann surface X, one chooses a disk $\varphi : U \to \mathbb{E}$, $\varphi(a) = 0$ around this point inside this neighborhood. Then f has order k at a if f_φ has order k at 0. It is clear that this definition doesn't depend on the choice of the disk.

5.3 Definition. *Let D be a divisor on a Riemann surface X. By a weak solution of D one understands a differentiable function without zero in the complement of the support of D, such that f has order $D(a)$ for each a of the support.*

Lemma 5.2 has then the obvious generalization:

5.4 Lemma. *Let f be a weak solution of a divisor D on a Riemann surface X and let g be a differentiable function on X with compact support. Then*

$$\frac{1}{2\pi \mathrm{i}} \int_X \frac{df}{f} \wedge dg = \sum_{a \in X} D(a)g(a).$$

(The sum is finite.)

Next we have to construct a weak solution. The essential part is a local construction:

5.5 Lemma. *Let a, b be points of the unit disk $\mathbb{E}$. There is a weak solution f of the divisor $(b) - (a)$ with the following additional property. There exists $0 < r < 1$ such that $f(z)$ is identical one for $r \leq |z| < 1$.*

Proof. In the case $a = b$ we can take $f \equiv 1$, hence we can assume $a \neq b$. We need a simple but basic observation from complex calculus. The function $(z-1)/(z+1)$ takes values on the negative real axis if and only if z is contained in the interval $[-1, 1]$. Hence the principal value of the logarithm defines a *holomorphic* function

$$\mathbb{C} - [-1, 1] \longrightarrow \mathbb{C}, \quad z \longmapsto \log \frac{z-1}{z+1}.$$

An easy consequence is:

Let a, b be two different points in the disk $|z| \leq r$. Then there exists in the complement $|z| > r$ a holomorphic branch of the logarithm $\log \frac{z-b}{z-a}$.

In our situation we can take $r < 1$. Now we consider a differentiable function ψ on $[r, 1]$ which is 1 close to r and 0 close to 1. The function

$$\exp\left(\psi \log \frac{z-b}{z-a}\right) \qquad r \leq |z| < 1$$

is 1 if $|z|$ is close to 1 and $\frac{z-a}{z-b}$ if $|z|$ is close to r. Hence we can glue it with the function

$$\frac{z-b}{z-a} \qquad |z| \leq r \ (z \neq a, b).$$

This gives the proof of Lemma 5.5. □

Again there is a generalization to Riemann surfaces:

5.6 Lemma. *Let $\alpha : [0, 1] \to X$ be a curve on a Riemann surface, $a = \alpha(0)$, $b = \alpha(1)$. There exists a weak solution f of the divisor $(b) - (a)$ with the following properties:*
1) *The function f is constant one outside some compact set.*
2) *For every closed differential ω on X one has*

$$\int_\alpha \omega = \frac{1}{2\pi i} \int_X \frac{df}{f} \wedge \omega.$$

Proof. First we mention that df vanishes outside a compact set. Hence the existence of the integral on the right hand side is only a local problem around the singularities a and b. This has been settled above.

Let $c = \alpha(t_0)$, $0 \leq t_0 < 1$ be another point on the curve. Assume that the problem has been solved for the curve $\alpha|[0, t_0]$. Denote the solution by f_1. Let similarly f_2 be a solution for the curve $\alpha|[t_0, 1]$. Then $f_1 f_2$ gives a solution for the total curve α. This observation allows us to restrict to the case, where

§5. Abel's theorem

the curve is contained in a disk U. In this disk we get a weak solution from Lemma 5.5. This solution can be extended by one to a solution on X. Now we have to prove the integral formula. In the disk U the differential ω is total, $\omega = dg$. Now we take a smaller disk $V \subset U$ whose closure is contained in U and such that f is one outside V. We can modify g to get a differentiable function with compact support such that the equation $\omega = dg$ still holds on V. Then

$$\int_\alpha \omega = g(b) - g(a)$$

and by Lemma 5.4 we also have

$$\frac{1}{2\pi i} \int_X \frac{df}{f} \wedge \omega = g(b) - g(a). \qquad \square$$

Now we can see that every divisor of degree zero on a compact Riemann surface admits a weak solution. We write the divisor in the form

$$D = (a_1) + \cdots + (a_n) - (b_1) - \cdots - (b_n).$$

We join a_i and b_i by a curve γ_i. Using Lemma 5.6 we find a weak solution f of the divisor D which has the additional property

$$\sum_{i=1}^n \int_{\alpha_i} \omega - \sum_{i=1}^n \int_{\beta_i} \omega = \frac{1}{2\pi i} \int_X \frac{df}{f} \wedge \omega$$

for all closed differentials ω. We will use this for *holomorphic* ω. Then

$$\frac{df}{f} \wedge \omega = \frac{\bar\partial f}{f} \wedge \omega.$$

Recall that f locally is of the form $z^k g(z)$ with a differentiable function g without zero. This implies (locally)

$$\frac{\bar\partial f}{f} = \frac{\bar\partial g}{g}.$$

Hence

$$\sigma := \frac{\bar\partial f}{f} \in \mathcal{A}^{0,1}(X)$$

is differentiable everywhere.

Let us assume now that the divisor D is in the kernel of the Abel–Jacobi map. Then we get

$$\frac{1}{2\pi i} \int_X \sigma \wedge \omega = \sum_{i=1}^n \int_{\gamma_i} \omega = 0$$

for all holomorphic ω. But this implies $\sigma = \bar\partial h$ (Corollary of Proposition 2.1). We use h to modify the weak solution f:

$$F = e^{-h} f.$$

We obtain

$$\bar\partial F = -e^{-g} f \bar\partial h + e^{-h} \bar\partial f = 0.$$

Hence F is a meromorphic solution. This gives Abel's Theorem.

5.7 Abel's theorem. *The homomorphism*

$$A : \mathrm{Pic}^0(X) \longrightarrow \mathrm{Jac}(X)$$

is injective.
In other words: A divisor of degree zero is principal if and only if its image in $\mathrm{Jac}(X)$ is zero.

As an example we treat a torus $X = \mathbb{C}/L$. Recall that $\Omega(X)$ is generated by "$\omega = dz$". We represent the divisor D of degree zero by a divisor $(a_1) + \cdots + (a_n) - (b_1) - \cdots - (b_n)$ in the interior of some fundamental parallelogram. We join a_i and b_i by curves γ_i inside this interior. Then

$$\sum_{i=1}^{n} \int_{\gamma_i} dz = \sum_{i=1}^{n} b_i - \sum_{i=1}^{n} a_i.$$

This gives the well-known Abel theorem for elliptic functions:
A divisor of degree zero $(a_1) + \cdots + (a_n) - (b_1) - \cdots - (b_n)$ on a torus is principal if and only if $a_1 + \cdots + a_n - b_1 - \cdots - b_n = 0$ (on the torus).

6. The Jacobi inversion problem

All what we have proved about the group $L \subset \Omega(X)^*$ is that it is a countable group. We want to prove more, namely that it is a lattice. By a lattice in a finite dimensional real vector space V we understand a subgroup L of the form

$$L = \mathbb{Z}e_1 + \cdots + \mathbb{Z}e_n$$

where $e_1, \ldots, e_n$ is a basis of V. A lattice is obviously a discrete subgroup which generates V as vector space. The converse is also true. We will use this without proof.

By a lattice in a finite dimensional complex vector space we understand a lattice of the underlying real vector space.

6.1 Proposition. *The set of periods $L \subset \Omega(X)^*$ of a compact Riemann surfaces is a lattice. Hence $\mathrm{Jac}(X)$ is a torus of real dimension $2g$.*

First part of the proof: L is discrete.
Let V be a complex vector space of dimension g. If $\mathfrak{M}$ is a set of sub-vector spaces whose intersection is 0, then there exist $\leq g$ vector spaces in $\mathfrak{M}$ whose intersection is zero. For a point $a \in X$ we can consider the subspace of all $\omega \in \Omega(X)$ which vanish at a. The above remark shows that there exist g points $a_1, \ldots, a_g$ such that a form $\omega \in \Omega(X)$ vanishes if it vanishes at these

§6. The Jacobi inversion problem

points. Let $\omega_1, \ldots, \omega_g$ be a basis of $\Omega(X)$. We choose disks $z_i : U_i \to \mathbb{E}$ around the points a_i. In these charts $\omega_i = f_{ij} dz_j$. Here we write f_{ij} as a function on U_i and not on $\mathbb{E}$. Then

$$A = \bigl(f_{ij}(a_j)\bigr)_{1 \leq i,j \leq g}$$

is an invertible matrix. In the disk U_i we define the holomorphic function

$$F_i(x) = \int_{a_i}^{x} \omega_i,$$

where the integral is taken along a path from a_j to x inside U_i. Then we consider the map

$$F : U_1 \times \cdots \times U_g \longrightarrow \mathbb{C}^g, \quad F(\xi_1, \ldots, \xi_g) = F_1(\xi_1) + \cdots + F_g(\xi_g).$$

The (complex) Jacobian of F with respect to the charts is (f_{ij}). From the theorem of invertible functions follows that

$$W := F(U_1 \times \cdots \times U_g) \subset \mathbb{C}^g$$

is a neighborhood of $F(a_1, \ldots, a_g)$.

Here we use a complex version of the theorem of invertible functions of several variables. But this is a consequence of the real version. The point is that F is differentiable in the real sense and the real functional determinant is the square of the absolute value of the complex functional determinant. To prove this, one has to express the real derivatives by the complex ones and then to use the Cauchy Riemann equations.

The basis $\omega_1, \ldots, \omega_g$ induces a dual basis $\omega_1^*, \ldots, \omega_g^*$ of $\Omega(X)^*$, namely

$$\omega_i^*(\omega_j) = \delta_{ij}.$$

We use this basis to identify the space $\Omega(X)^*$ with $\mathbb{C}^g$. Then the map F can be read as a map

$$F : U_1 \times \cdots \times U_g \longrightarrow \Omega(X)^*.$$

The advantage is that it doesn't depend on the choice of the basis, since

$$F(\xi_1, \ldots, \xi_g)(\omega) = \sum_{i=1}^{g} \int_{a_i}^{\xi_i} \omega.$$

(We used the coordinates just since the theorem of invertible functions often is not formulated in a coordinate invariant manner.)

We claim that W has empty intersection with L. (Since L is a group, this implies the discreteness of L.) We argue by contradiction and assume

$F(\xi_1, \ldots, \xi_g) = 0$ Since $F(\xi_1, \ldots, \xi_g)$ represents the image of the divisor $D = (\xi_1) + \cdots + (\xi_g) - (a_1) - \cdots - (a_g)$ under the Abel–Jacobi map, we can apply Abels's theorem 5.7. There exists a meromorphic function f with $(f) = D$. The function f has poles of order one in the a_i. We denote by $C_i \neq 0$ the residue of f with respect to the chart z_i. Then

$$\operatorname{Res}_{a_i}(f\omega_i) = C_i f_{ij}(a_j).$$

From the residue theorem applied to the meromorphic differentials $f\omega_i$ follows

$$\sum_{j=1}^{g} C_j \varphi_{ij}(a_j) = 0.$$

But then the matrix A cannot be invertible, which gives a contradiction.

Second part of the proof: L *generates* $\Omega(X)^*$ *as real vector space.*

We have to show that a real linear form on $\Omega(X)^*$ that vanishes on L is zero. Every real linear form is the real part of a $\mathbb{C}$-linear (complex) linear form. Every complex linear form on $\Omega(X)^*$ is of the form

$$l \longmapsto l(\omega)$$

for some $\omega \in \Omega(X)$. The real part of this linear form vanishes on L if and only if the real parts of the periods of ω are zero. But then ω is zero (Proposition 4.4). □

The first part of the proof of Proposition 6.1 shows that a full neighbourhood of $0 \in \operatorname{Jac}(X)$ is contained in the image of the Abel–Jacobi map. The image is also a group. Obviously a torus is generated as group by any neighborhood of the origin. Hence we obtain that the Abel-Jacbobi map is not only injective but also surjective.

6.2 Theorem. *The Abel–Jacobi map*

$$A : \operatorname{Pic}^0(X) \xrightarrow{\sim} \operatorname{Jac}(X)$$

is an isomorphism.

We come back to the Abel–Jacobi map in the form

$$A : X \longrightarrow \operatorname{Jac}(X).$$

It depends on the choice of a base point a.

6.3 Remark. *Let X be compact Riemann surface of genus $g > 1$. Then*

$$A : X \longrightarrow \operatorname{Jac}(X)$$

is injective.

§6. The Jacobi inversion problem

Proof. We mention that a bijective holomorphic map $f : X \to Y$ between Riemann surfaces is biholomorphic. This is known from complex calculus in the case of open domains in $\mathbb{C}$. The general case works in the same way. Let f be a meromorphic function on a compact Riemann surface with divisor $(b) - (a)$. Then it is of order one and hence defines a bijective map $X \to \bar{\mathbb{C}}$. Hence X is biholomorphic equivalent to $\bar{\mathbb{C}}$. This implies Remark 6.3. □

We call a point $a \in X^g$ *generic* if every holomorphic differential ω, which vanishes at all points occurring in a, vanishes identically. In the fist part of the proof of Proposition 6.1 we have shown that there are generic points. The argument shows a little more, namely:

6.4 Remark. *The set of generic points is open and dense in X^g. The map $A : X^g \longrightarrow \mathrm{Jac}(X)$ is locally topological on the set of generic points.*

Let now X be a compact Riemann surface of genus one. Then $\mathrm{Jac}(X) = \mathbb{C}/L$ is also a Riemann surface. The map $A : X \to \mathrm{Jac}(X)$ is injective. Since it is also surjective we obtain the following theorem.

6.5 Theorem. *Every compact Riemann surface of genus one is biholomorphic equivalent to a torus $\mathbb{C}/L$.*

Now we switch to $A : X^d \to \mathrm{Jac}(X)$. One may ask, whether this map is surjective for suitable d. Since the real dimension of X^d is $2d$ and that of $\mathrm{Jac}(X)$ is $2g$, one can hope that this is true for $d = g$.

6.6 Theorem. *The Abel–Jacobi map*

$$A : X^g \longrightarrow \mathrm{Jac}(X)$$

is surjective.

Proof. let $a \in X$ be the base point. Since $\mathrm{Pic}^0(X) \to \mathrm{Jac}(X)$ is surjective, we only must show the following:

Every divisor D of degree 0 is quivalent to a divisor $(b_1) + \cdots + (b_g) - g \cdot (a)$.

For the proof we consider the divisor $D' = D + g(a)$. It has degree g. Riemann–Roch implies that $\dim \mathcal{O}_D(X) \geq 1$. Hence there exists a non-zero meromorphic function f such that $(f) + D' \geq 0$. Since the degree of this divisor is g we get $(f) + D' = (b_1) + \cdots + (b_g)$. This gives $(f) + D = (b_1) + \cdots + (b_g) - g(a)$. □

The symmetric group S_g acts on X^g by permutation of the components. The quotient

$$X^{(g)} = X^g / S_g$$

can be considered as the the set of unordered n-tuples of X. Since the map $X^g \to \mathrm{Jac}(X)$ does not depend on the ordering of the points, it factors through the natural projection $X^g \to X^{(g)}$. We obtain the *Jacobi map*

$$J : X^{(g)} \longrightarrow \mathrm{Jac}(X).$$

The Jacobi inversion problem asks for the inversion of this map. Actually for $g > 1$ this map is not bijective but it is close to a bijective map. The correct statement is:

6.7 Jacobi inversion theorem, unprecise statement. *The map*
$$J : X^{(g)} \longrightarrow \mathrm{Jac}(X).$$
*is **bimeromorphic**.*

Up to no we did not define what bimeromorphic means. This will be part of what follows.

7. The fibres of the Jacobi map

We can identify the elements of $X^{(g)}$ with divisors $D \geq 0$ of degree g. We have to determine the fibre $J^{-1}(J(D))$. It consists of all divisors $D' \geq 0$ of degree g which are equivalent to D. Then there exist a meromorphic function with $D' = D + (f)$. Because of $D' \geq 0$ we have $f \in \mathcal{O}_D(X)$. The function f is determined up to a constant factor. Hence it is better to identify two f if they are contained in the same one-dimensional complex sub-vector space. Recall that the set of all one-dimensional sub-vector spaces of a vector space V is called the projective space $P(V)$. There is a natural map $V - \{0\} \to P(V)$. We use this (for finite dimensional V) to define a topology on $P(V)$, the quotient topology. It is easy to show that this is a compact space.

7.1 Remark. *Let $D \in X^{(g)}$. There is a natural bijective map*
$$P(\mathcal{O}_D(X)) \xrightarrow{\sim} f^{-1}(f(D)), \quad f \longmapsto D + (f).$$

The space X^g carries the product topology and $X^{(g)} = X^g / S_g$ carries the quotient topology. It is easy to see that this is a Hausdorff space.

7.2 Lemma. *The map*
$$P(\mathcal{O}_D(X)) \xrightarrow{\sim} f^{-1}(f(D)), \quad f \longmapsto D + (f),$$
is topological.

Corollary. *The fibres of the Jacobi map are connected.*

It is sufficient to show that the map is continuous. We omit the proof. □

From Remark 6.4 we know that a generic point in X^g is isolated in the fibre of $X^g \to \mathrm{Jac}(X)$. As a consequence its image in $X^{(g)}$ is isolated in its fibre as well. From Lemma 7.2 follows that it is the only point in its fibre. Hence we obtain the following theorem.

7.3 Theorem. *The Jacobi map $J : X^{(g)} \to \operatorname{Jac}(X)$ is surjective. There exists an open and dense subset of $X^{(g)}$ on which J is injective.*

This can be considered as a weak version of the Jacobi inversion theorem, which states that J is bimeromorphic.

8. The Jacobian inversion problems and abelian functions

We need some basic facts about holomorphic functions in several variables. A function $f : D \to \mathbb{C}^m$, where $D \subset \mathbb{C}^n$ is an open subset, is called complex differentiable at a point $a \in D$ if there exists a $\mathbb{C}$-linear map $J(f,a) : \mathbb{C}^n \to \mathbb{C}^m$ such that

$$f(z) - f(a) = J(f,a)(z-a) + r(z), \qquad \lim_{z \to a} \frac{r(z)}{||z-a||} = 0.$$

Hence "complex differentiable" implies "real differentiable". The usual permanent properties including the theorem of invertible functions carry over to the complex case. The function f is called holomorphic if and only if it is complex differentiable at each point. This is the case if and only if every component of f is holomorphic. There is a basic lemma.

8.1 Lemma. *Let $f : D \to \mathbb{C}$ be a continuous function on an open domain $D \subset \mathbb{C}^n$ such that f is holomorphic in each variable separately. Then f is holomorphic. Moreover for every point $a \in D$ there exists a power series, which converges absolutely and locally uniform in some ball $U_r(a) \subset D$ and represents there f:*

$$f(z) = \sum_{\nu \in \mathbb{N}_0^n} (z-a)^\nu \qquad (a \in U_r(a)).$$

Here we use the usual writing with multiindices. The proof uses a simple generalization of the Cauchy integral formula. Applying the usual one in each variable one obtains

$$f(z) = \frac{1}{(2\pi i)^n} \oint_{|\zeta_1 - a_1| = r} \cdots \oint_{|\zeta_n - a_n| = r} \frac{f(\zeta) d\zeta_1 \ldots d\zeta_n}{(z_1 - \zeta_1) \cdots (z_n - \zeta_n)}.$$

Now the same argument as in the case $n = 1$ works. One expands the integrand in to a (multivariable) geometric series and interchanges summation and integration.

The power series expansion shows a weak form of the principle of analytic continuation: if two analytic functions $f, g : D \to \mathbb{C}^m$ on a connected open subset $D \subset \mathbb{C}^n$ agree on some non empty open subset, then they agree everywhere.

Analytic manifolds

Analytic manifolds of complex dimension $n \geq 1$ are the straight forward generalization of Riemann surfaces ($n = 1$). Hence we can keep short:

An analytic manifold is a geometric space, which is locally isomorphic to a space $(U, \mathcal{O}_U)$, where $U \subset \mathbb{C}^n$ is an open subset and $\mathcal{O}_U$ is the sheaf of holomorphic functions. An analytic map between analytic manifolds is just a morphism of geometric spaces.

Basic facts of Riemann surfaces carry over to analytic manifolds. We mention just the following weak form of the principle of analytic continuation which easily follows by means of power series:

Let $f, g : X \to Y$ be two analytic maps between to analytic manifolds. Assume that X is connected and that f and g agree on some non-empty open subset. Then f and g agree on the whole X.

Meromorphic functions

Zero sets of analytic functions are not discrete in general as for example the function $z_1 \cdot z_2$ on $\mathbb{C}^2$ shows. This makes the notion of a meromorphic function more delicate in the higher dimensional case. An example of a meromorphic function on $\mathbb{C}^2$ should be z_1/z_2. There is no clean way to define a value for this function at the origin 0. Hence it would be false to define meromorphic functions simply as holomorphic maps into $\bar{\mathbb{C}}$. This works only in the case $n = 1$. We proceed as follows:

Let X be an analytic manifold. We consider pairs (U, f), where $U \subset X$ is an open an dense subset of X and $f : U \to \mathbb{C}$ is an analytic function. We call this pair meromorphic on X, if for every point $a \in X$ (only $a \notin U$ is of interest), there exists a small open connected neighborhood U and two analytic functions $g, h : U \to \mathbb{C}$, such that h is not identically zero and such that

$$f(x) = \frac{g(x)}{h(x)} \quad \text{for all} \quad x \in U, \ h(x) \neq 0.$$

Two meromorphic pairs (U, f) and (V, g) are called equivalent if f and g agree on $U \cap V$. A *meromorphic function* on X is a full equivalence class of such pairs. It is easy to define the sum and product of two meromorphic functions. It is also not difficult to prove that for connected X the set of all meromorphic functions is a field. (We don't know that U is connected. But this does not matter, since the condition of meromorphicity concerns the whole domain D.) We denote this field by $K(X)$.

Examples of analytic manifolds

The first example is a torus $X_L := \mathbb{C}^n/L$, where L is a lattice. As in the case $n = 1$ the meromorphic functions on X_L correspond uniquely to the L-periodic meromorphic functions on $\mathbb{C}^n$. Such functions are called "*abelian functions*". They generalize the elliptic functions.

§8. The Jacobian inversion problems and abelian functions

The direct product $X \times Y$ of two analytic manifolds carries also a structure as analytic manifold. As a consequence, the power $X^n = X \times \ldots \times X$ of a Riemann surface is an analytic manifold.

We go back to the Abel–Jacobi map

$$A : X^g \longrightarrow \mathrm{Jac}(X).$$

Pulling back a function we get an imbedding of fields

$$K(\mathrm{Jac}(X)) \longrightarrow K(X^g).$$

The image is contained in the subfield $K(X^g)^{S_g}$ of all functions which are invariant under arbitrary permutations.

It can be shown that the symmetric power $X^{(g)}$ is an analytic manifold as well and that $K(X^{(g)})$ can be naturally identified with $K(X^g)^{S_g}$. We will not prove this and take $K(X^g)^{S_g}$ just as substitute for the correct field $K(X^{(g)})$.

8.2 Jacobi inversion theorem, precise statement. *The natural map*

$$K(\mathrm{Jac}(X)) \longrightarrow K(X^g)^{S_g}$$

is an isomorphism of fields.

We will not prove this here. An elementary proof can be found in [Fr1], Theorem VI.13.13.

We work out a special case of Theorem 8.2 which will lead us to a formulation which is close to what Jacobi had in mind. Let $f : X \to \bar{\mathbb{C}}$ be a non-constant meromorphic function. It induces an analytic map

$$f^g : X^g \to \bar{\mathbb{C}}^g.$$

The projections give g analytic maps

$$f_\nu : X^g \longrightarrow \bar{\mathbb{C}} \qquad (1 \leq \nu \leq g).$$

They are just defined by

$$f_\nu(\xi_1, \ldots, \xi_g) = f(\xi_\nu).$$

They can be considered as meromorphic functions on X^g. But they are not invariant under S_g. Hence we consider the elementary symmetric expressions

$$E_k = \sum_{1 \leq \nu_1 \leq \ldots \leq \nu_k \leq g} f(\xi_{\nu_1}) \ldots f(\xi_{\nu_k})$$

They are also meromorphic functions on X^g with the advantage to be symmetric. The Jacobi inversion theorem predicts the *existence of abelian functions* F_ν on X_L whose pull-back are the E_k.

Historical cases

We consider the Riemann surface X of the function $\sqrt{P(z)}$, where P is a polynomial of degree 3 or 4 without multiple zero. Recall that the corresponding Riemann surface has genus one. Up to a finite number of points it is the curve (z, w), $w^2 = P(z)$. The natural projection $(z, w) \mapsto z$ is a meromorphic function on X, which we take for f. Recall that $dz/\sqrt{P(z)}$ generates $\Omega(X)$. The Abel–Jacobi map is

$$X \longrightarrow \mathbb{C}/L, \quad x \longmapsto \int_a^x \frac{dz}{\sqrt{P(z)}}.$$

The inversion theorem says in this case that the function z is the pull-back of an elliptic function φ. "Pull-back" means that the composition of φ with the Abel–Jacobi map gives x, i.e.

$$x = \varphi\left(\int_a^x \frac{dz}{\sqrt{P(z)}}\right).$$

This reflects the classical observation of Abel that the inversion of the elliptic integral gives an elliptic function.

The big question was how to generalize this to the hyperelliptic case. Let us consider the case $\sqrt{P(z)}$, where P now is of degree 5 ore 6 without multiple zero. Recall that a basis of $\Omega(X)$ in this case is given by

$$\frac{dz}{\sqrt{P(z)}}, \quad \frac{z\,dz}{\sqrt{P(z)}}.$$

The Abel–Jacobi map now is given by

$$(\xi_1, \xi_2) \longmapsto \left(\int_a^{\xi_1} \frac{dz}{\sqrt{P(z)}} + \int_a^{\xi_2} \frac{dz}{\sqrt{P(z)}}, \int_a^{\xi_1} \frac{z\,dz}{\sqrt{P(z)}} + \int_a^{\xi_2} \frac{z\,dz}{\sqrt{P(z)}}\right)$$

Now we have to consider the two elementary symmetric functions $\xi_1 + \xi_2$, $\xi_1\xi_2$. The Jacobi inversion theorem states:

There are two abelian functions φ_1, φ_2 such that

$$\xi_1 + \xi_2 = \varphi_1\left(\int_a^{\xi_1} \frac{dz}{\sqrt{P(z)}} + \int_a^{\xi_2} \frac{dz}{\sqrt{P(z)}}, \int_a^{\xi_1} \frac{z\,dz}{\sqrt{P(z)}} + \int_a^{\xi_2} \frac{z\,dz}{\sqrt{P(z)}}\right)$$

$$\xi_1\xi_2 = \varphi_2\left(\int_a^{\xi_1} \frac{dz}{\sqrt{P(z)}} + \int_a^{\xi_2} \frac{dz}{\sqrt{P(z)}}, \int_a^{\xi_1} \frac{z\,dz}{\sqrt{P(z)}} + \int_a^{\xi_2} \frac{z\,dz}{\sqrt{P(z)}}\right)$$

Jacobi formulated this as a problem before the theoy of Riemann surfaces existed. The theory of Riemann surfaces enabled to proof this and moreover, to reformulate it in a precise and very natural way. Moreover Jacobi's inversion problem opened the door to the theory of complex analytic functions in several variables.

Chapter VIII. Dimension formulae for automorphic forms

In his paper [Bo1], at the end of Sect. 4, Borcherds mentions without proof a beautiful formula for dimensions of spaces of vector valued modular forms of weight ≥ 2 with respect to the full modular group (here reproduced as Theorem 7.1).

Skoruppa informed me that he derived already 1985 in his Ph.D. thesis [Sk] such dimension formulas for all weights by means of the Shimura trace formula.

More general results, including arbitrary Fuchsian groups, can be found in the paper [Bo2] of Borcherds, Sect. 7. Most of them have been proved by the Selberg trace formula, see [Iv] and also [Fi]. The Selberg trace formula in its standard form causes the restriction that the weight is > 2. Borcherds mentions that "with a bit more care this also works for weight 2". As we mentioned, this bit more care was taken in important cases already 1985 in the thesis of Skoruppa.

The purpose of this chapter is to produce the dimension formula in all weights, for general Fuchsian groups and for arbitrary real weights.

We consider arbitrary discrete subgroups $\Gamma \subset \mathrm{SL}(2,\mathbb{R})/\pm$ with finite volume of the fundamental domain. For them we consider vector valued modular forms of a real weight r. They have the transformation property

$$f(\gamma\tau) = \gamma'(\tau)^{-r/2} v(\gamma) f(\tau),$$

where $v(\gamma)$ is matrix-valued. We assume that all $v(\gamma)$ are unitary.

In a first approach we assume a little more, namely that r is rational, that the $v(\gamma)$ are of finite order, and that they can be diagonalized simultaneously for all γ in a subgroup of finite index Γ_0. The precise form is formulated in Assumption 4.1. These extra assumptions are not really necessary. But we found that these restrictions are convenient and they cover all cases which occur usually in the theory of modular forms. For sake of completeness we treat the most general case in an Appendix.

Since we allow non-integral weights, there is an ambiguity in the definition of $\gamma'(\tau)^{-r/2}$. One standard way to overcome this, is, to use covering groups of $\mathrm{SL}(2,\mathbb{R})$. Instead of this we found it convenient to use the old-fashioned method of multiplier systems, here in a matrix-valued sense. They could be called also "projective representations". In Sect. 6 we reformulate the results for representations of the two fold metaplectic covering of $\mathrm{SL}(2,\mathbb{Z})$ and reproduce Borcherds' formula in [Bo1].

Our proof rests on the Riemann–Roch formula for vector bundles (and not on the Selberg trace formula). The idea is to use a sufficiently small normal subgroup Γ_0 for which the vector bundle splits into a sum of line bundles. This gives a reduction to the well-known case of scalar valued modular forms.

1. Fuchsian groups and Riemann surfaces

The group $\mathrm{SL}(2,\mathbb{R})$ can be considered as a closed subset of $\mathbb{R}^4$. We equip it with the induced topology. It is a locally compact space with countable basis of the topology. We consider discrete subgroups $\Gamma \subset \mathrm{SL}(2,\mathbb{R})$. Here a subset S of a topological space X is called a discrete subset if it is closed and if the induced topology is the discrete topology (every set is open). For locally compact X this means that the intersection of S with each compact subset of X is finite. A basic example of a discrete subgroup in $\mathrm{SL}(2,\mathbb{R})$ is the group $\mathrm{SL}(2,\mathbb{Z})$.

We recall that each matrix $M \in \mathrm{GL}(2,\mathbb{C})$ induces a biholomorphic transformation of the Riemann sphere onto itself:

$$M\tau = (a\tau + b)(c\tau + d)^{-1}, \quad M = \begin{pmatrix} a & b \\ c & d \end{pmatrix}.$$

If M is real and has positive determinant, then M maps $\mathbb{H}$ onto itself and it maps also the extended real line $\mathbb{R} \cup \{\infty\}$ onto itself. In particular, the group $\mathrm{SL}(2,\mathbb{R})$ acts on the upper half plane. Two elements of $\mathrm{SL}(2,\mathbb{R})$ define the same transformation if and only if they differ by a sign. The elements $\pm E$ (E denotes the unit matrix) build a normal subgroup of $\mathrm{SL}(2,\mathbb{R})$. Hence, in the following, we are more interested in the factor group $\mathrm{SL}(2,\mathbb{R})/\pm$. But at the moment we prefer to work with matrices.

1.1 Definition. *An element $M \in \mathrm{SL}(2,\mathbb{R})$ is called **elliptic** if $|\mathrm{tr}(M)| < 2$ and **parabolic** if it is different from $\pm E$ and if $|\mathrm{tr}(M)| = 2$.*

From the fixed point equation

$$M(\tau) = \tau \iff c\tau^2 + (d-a)\tau - b = 0$$

we obtain the following result.

1.2 Lemma. *An element $M \in \mathrm{SL}(2,\mathbb{R})$ is elliptic if and only if it is different from $\pm E$ and if it has a fixed point in $\mathbb{H}$. This fixed point is unique.*

An element $M \in \mathrm{SL}(2,\mathbb{R})$ is parabolic if and only if it has a unique fixed point in $\mathbb{R} \cup \{\infty\}$.

§1. Fuchsian groups and Riemann surfaces

We say that a subgroup $\Gamma \subset \mathrm{SL}(2,\mathbb{R})$ acts *discontinuously* if for each two compact subsets $K_1, K_2 \subset \mathbb{H}$ the set

$$\{M \in \Gamma; \quad M(K_1) \cap K_2 \neq \emptyset\}$$

is finite. It is enough to demand this for $K_1 = K_2$ (consider $K_1 \cup K_2$). In particular, the stabilizer

$$\Gamma_a := \{M \in \Gamma; \quad Ma = a\}$$

is a finite subgroup for every $a \in \mathbb{H}$.

1.3 Proposition. *Each discrete subgroup $\Gamma \subset \mathrm{SL}(2,\mathbb{R})$ acts discontinuously.*

The proof of this Proposition and its converse can be found in [Fr1], III.2.3 or [Fr2], I.1.2. Here we just mention that this is an immediate consequence of the fact that the map

$$\mathrm{SL}(2,\mathbb{R}) \longrightarrow \mathbb{H}, \quad M \longmapsto M(\mathrm{i}),$$

is proper. This follows from the fact that the stabilizer of i is the special orthogonal group $\mathrm{SO}(2,\mathbb{R})$ which is a compact subgroup.

Two points $a, b \in \mathbb{H}$ are called equivalent with respect to Γ if there exists $M \in \Gamma$ such that $b = Ma$. It is easy to check that this is an equivalence relation. We denote by $[a]$ the equivalence class of a and we denote by $\mathbb{H}/\Gamma$ the set of all equivalence classes. We equip $\mathbb{H}/\Gamma$ with the quotient topology. There is a natural projection $\mathbb{H} \to \mathbb{H}/\Gamma$. It is continuous and open (i.e the images of open sets are open).

1.4 Proposition. *Let $\Gamma \subset \mathrm{SL}(2,\mathbb{R})$ be a discrete subgroup. The quotient $\mathbb{H}/\Gamma$ is a (Hausdorff) locally compact space with countable basis of the topology. Moreover the following holds.*

For each $a \in \mathbb{H}$ there exists an open neighborhood $U(a)$ which is invariant under the stabilizer Γ_a and such that two points in $U(a)$ are equivalent mod Γ if and only if they are equivalent with respect to Γ_a.

For a proof we refer to [Fr2], I.1.7. □

We can define the quotient space $U(a)/\Gamma_a$ in the obvious way. There is a natural map

$$U(a)/\Gamma_a \longrightarrow \mathbb{H}/\Gamma.$$

The condition in Proposition 1.4 says that this map is injective. Moreover, it is an open embedding. This means that its image is open and that this map induces a topological map onto this image. Hence $\mathbb{H}/\Gamma$ looks locally (close to $[a]$) like $\mathbb{H}/\Gamma_a$.

Extension by cusps

Let $\Gamma \subset \mathrm{SL}(2,\mathbb{R})$ be a discrete subgroup. An element $a \in \mathbb{R} \cup \{\infty\}$ is called a *cusp* of Γ if it is the fixed point of a parabolic element $M \in \Gamma$. The elements of $\mathrm{SL}(2,\mathbb{R})$ which fix ∞ are of the form

$$\begin{pmatrix} a & b \\ 0 & a^{-1} \end{pmatrix}.$$

Only in the case $a = \pm 1$ they are parabolic. These elements are called translation matrices.

1.5 Lemma. *Assume that the discrete group Γ has cusp ∞. Then each element of the stabilizer Γ_∞ is a translation matrix. The image of Γ_∞ in $\mathrm{SL}(2,\mathbb{R})/\pm$ is an infinite cyclic group.*

Proof. The formula

$$\begin{pmatrix} a & * \\ 0 & a^{-1} \end{pmatrix}^n \begin{pmatrix} 1 & b \\ 0 & 1 \end{pmatrix} \begin{pmatrix} a & * \\ 0 & a^{-1} \end{pmatrix}^{-n} = \begin{pmatrix} 1 & a^{2n}b \\ 0 & 1 \end{pmatrix}$$

shows that Γ_∞ would have the unit matrix as accumulation point if it would contain a non trivial translation and an element with $a \neq \pm 1$. Hence all elements of Γ_∞ act as translations $\tau \mapsto \tau + b$. There exists a smallest positive b. Then every transformation of Γ_∞ acts as $\tau \mapsto \tau + nb$ with some integer n. This means that the image of the group Γ_∞ in $\mathrm{SL}(2,\mathbb{R})/\pm$ is a cyclic group generated by $\pm\begin{pmatrix} 1 & b \\ 0 & 1 \end{pmatrix}$. □

Let κ be an arbitrary cusp which is different from ∞. Then we can choose a matrix $N \in \mathrm{SL}(2,\mathbb{R})$ such that $N(\kappa) = \infty$. For example, one can take

$$N = \begin{pmatrix} 0 & 1 \\ -1 & \kappa \end{pmatrix}.$$

Then the conjugated group $N\Gamma N^{-1}$ has cusp ∞. This transformation is often used to reduce properties of arbitrary cusps to the cusp ∞. We have to consider the stabilizer Γ_κ of a cusp. We have

$$N\Gamma_\kappa N^{-1} = (N\Gamma N^{-1})_\infty.$$

Hence transforming κ to ∞ allows to restrict to $\kappa = \infty$.

In the following, we denote by $\mathbb{H}^*$ the union of $\mathbb{H}$ and the set of cusps. Of course this depends on Γ. It is easy to see that a subgroup $\Gamma_0 \subset \Gamma$ of finite index has the same cusps as Γ and hence leads to the same $\mathbb{H}^*$. The group Γ acts on $\mathbb{H}^*$.

We want to define a topology on $\mathbb{H}^*$. This will not be the induced topology of the Riemann sphere. Recall that in the Riemann sphere a sequence (a_n)

§1. Fuchsian groups and Riemann surfaces

tends to ∞ if the absolute values $|a_n|$ tend to ∞. But we want to have the following: a sequence (a_n) of points in the upper half plane should tend to ∞ if the imaginary parts tend to ∞. Hence a typical neighborhood of ∞ should be defined by $\operatorname{Im} \tau > C$ (and not by $|\tau| > C$ as in the topology of the Riemann sphere). If we apply to the set $\operatorname{Im} \tau > C$ a transformation $M \in \operatorname{SL}(2, \mathbb{R})$ which fixes ∞ then we obtain a set of the same form (usually with different C). In all other cases we obtain an open disk in the upper half plane which is tangent to the real axis.

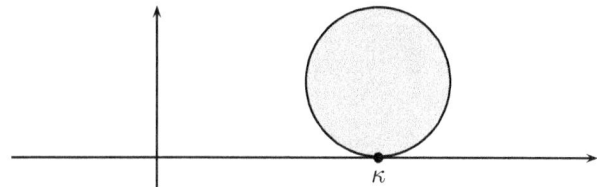

By a *horocycle* at the cusp κ we understand such a disk if $\kappa \neq \infty$ or a set of the form $\operatorname{Im} \tau > C > 0$ otherwise.

1.6 Definition. *Let $\Gamma \subset \operatorname{SL}(2, \mathbb{R})$ be a discrete subgroup. A subset $U \subset \mathbb{H}^*$ is called open if the following two conditions hold.*

a) *The intersection $U \cap \mathbb{H}$ is open in the usual sense.*
b) *Assume that the cusp κ is contained in U. Then U contains a horocycle at the cusp κ.*

Hence a typical neighborhood of a cusp κ is the union of a horocycle at κ and $\{\kappa\}$. It is trivial that this defines a topology on $\mathbb{H}^*$. It is rather clear that $\mathbb{H}$ is an open subset of $\mathbb{H}^*$ and that the induced topology is the usual one.

But this topology has some properties which look rather strange at a first glance. For example, each cusp has a neighborhood which contains no other cusp besides κ. This means that the set of cusps is a discrete subset of $\mathbb{H}^*$. We also mention that $\mathbb{H}^*$ is usually not locally compact.

A basic example is the group $\operatorname{SL}(2, \mathbb{Z})$. It is easy to see that all rational numbers and ∞ are cusps. Hence $\mathbb{H}^* = \mathbb{H} \cup \mathbb{Q} \cup \{\infty\}$ in this case and $\mathbb{Q}$ is a discrete subset!

Every element $M \in \Gamma$ induces a topological map from $\mathbb{H}^*$ onto itself. It is clear how to define the quotient space $X_\Gamma := \mathbb{H}^*/\Gamma$ and the natural projection

$$\mathbb{H}^* \longrightarrow X_\Gamma, \quad a \longmapsto [a].$$

It is continuous and open. The following generalization of Proposition 1.4 holds.

1.7 Proposition. *Let $\Gamma \subset \operatorname{SL}(2, \mathbb{R})$ be a discrete subgroup. The quotient $X_\Gamma = \mathbb{H}^*/\Gamma$ is a (Hausdorff) locally compact space with countable basis of the topology. Moreover, the following holds.*

For each $a \in \mathbb{H}^*$ there exists an open neighborhood $U(a)$ which is invariant under the stabilizer Γ_a and such that two points in $U(a)$ are equivalent mod Γ if and only if they are equivalent with respect to Γ_a.

The set of cusp classes is discrete in X_Γ.

A proof can be found in [Fr2], I.1.13. We just mention that the most involved part of the proof is the Hausdorff property of X_Γ. If Γ is a subgroup of finite index in $\mathrm{SL}(2,\mathbb{Z})$, the proof is easier. We refer to [Fr1], Proposition IV.14.6, for readers who are interested only in this case. □

We define a geometric structure $\mathcal{O}$ on X_Γ. Let $U \subset X_\Gamma$ be an open subset. We denote by $\tilde{U}$ the inverse image of U in $\mathbb{H}$. By definition, a function $f : U \to \mathbb{C}$ belongs to $\mathcal{O}$ if it is continuous and if the composition of the natural projection $\tilde{U} \to U$ and f is a holomorphic function $\tilde{U} \to \mathbb{C}$ in the usual sense.

1.8 Proposition. *The geometric structure $\mathcal{O}$ defines a structure as Riemann surface on X_Γ.*

Proof. We have to construct for each point of X_Γ an open neighborhood U, such the geometric space $(U, \mathcal{O}|U)$ is isomorphic to an open subset $V \subset \mathbb{C}$, equipped with the usual holomorphic structure. Let $a \in \mathbb{H}^*$ be a representative of this point. We choose an open neighborhood $U(a)$ which is invariant under Γ_a and such that $U(a)/\Gamma_a \to X_\Gamma$ is an open embedding. We equip $U(a)/\Gamma_a$ in the same way with a structure as geometric space as we did for X_Γ. Then the open embedding induces an isomorphism of the geometric space $U(a)/\Gamma_a$ onto its image in X_Γ (equipped with the restricted structure). Hence it is sufficient to show that $U(a)/\Gamma_a$ is a Riemann surface.

First we treat the case that a is not a cusp. We consider the biholomorphic map from $\mathbb{H}$ onto the unit disk $\mathbb{E}$

$$\alpha : \mathbb{H} \xrightarrow{\sim} \mathbb{E}, \quad \tau \longmapsto w = \frac{\tau - a}{\tau - \bar{a}}.$$

Every element $M \in \Gamma_a$ induces a transformation of the unit disk

$$\mathbb{E} \longrightarrow \mathbb{E}, \quad w \longmapsto \alpha(M(\alpha^{-1}(w))).$$

This transformation fixes the origin. By a well-known result of complex analysis it is of the form $w \mapsto \zeta w$ where $|\zeta| = 1$. Since Γ_a is finite we obtain a finite subgroup of the multiplicative group of complex numbers. Let n be its order. Another well-known result says that this group is the set of all nth roots of unity

$$G := \{e^{2\pi i \nu / n}; \ 0 \le \nu < n\}.$$

The map α induces a topological map

$$\mathbb{H}/\Gamma_a \xrightarrow{\sim} \mathbb{E}/G.$$

§2. Vector valued automorphic forms

We consider the natural projection $\mathbb{E} \to \mathbb{E}/G$ and compare it with the map $\mathbb{E} \to \mathbb{E}, w \mapsto w^n$. There exists a unique bijective map $\varphi : \mathbb{E}/G \to \mathbb{E}$ such the diagram

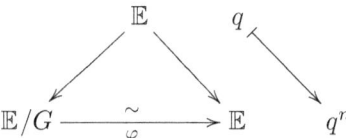

commutes. A trivial result of complex analysis says that a function f on the unit disk is holomorphic if and only if $f(w^n)$ is holomorphic. This shows that the bijection $\varphi : \mathbb{E}/G \to \mathbb{E}$ is an isomorphism of geometric spaces if one equips $\mathbb{E}$ with the usual holomorphic structure. This shows that $\mathbb{E}/G$ and hence $\mathbb{H}/\Gamma_a$ is a Riemann surface.

The case of a cusp a is similar. We can assume that a is the cusp ∞. We have to show that $(\mathbb{H} \cup \{\infty\})/\Gamma_\infty$ is a Riemann surface. Again, we construct an isomorphism $\varphi : (\mathbb{H} \cup \{\infty\})/\Gamma_a \to \mathbb{E}$ of geometric spaces. Recall that there exists a real number $b \neq 0$ that Γ_a acts be the translations $\tau \mapsto \tau + nb$, $n \in \mathbb{Z}$. We take for φ the map

$$\varphi([\tau]) = e^{2\pi i \tau/b} \qquad (= 0 \text{ for } \tau = \infty).$$

The topology of X_Γ has been defined in such a way that this map is topological. Moreover it is an isomorphism of geometric spaces. This shows that $(\mathbb{H} \cup \{\infty\})/\Gamma_\infty$ is a Riemann surface. □

A point $a \in \mathbb{H}$ is called an *elliptic fixed point* of Γ if it is the fixed point of an elliptic element.

2. Vector valued automorphic forms

In the following, we will work with the group $\mathrm{Aut}(\mathbb{H})$ of biholomorphic transformations of the upper half plane. As we explained, there is a natural homomorphism $\mathrm{SL}(2, \mathbb{R}) \to \mathrm{Aut}(\mathbb{H})$. It is well-known that this is surjective. Hence we get an isomorphism of groups

$$\mathrm{SL}(2, \mathbb{R})/\pm \xrightarrow{\sim} \mathrm{Aut}(\mathbb{H}).$$

In the following we will prefer to work with this group and not with $\mathrm{SL}(2, \mathbb{R})$. Hence Γ will denote in the following a subgroup of $\mathrm{Aut}(\mathbb{H})$. Each element of $\mathrm{Aut}(\mathbb{H})$ determines a matrix in $\mathrm{SL}(2, \mathbb{R})$ up to the sign.

Let $\alpha : D \to D'$ be a biholomorphic mapping between two domains in the complex plane. Once for ever we choose a holomorphic logarithm $\log \alpha'(\tau)$ and define then

$$j_r(\alpha, \tau) := \alpha'(\tau)^{-r/2} := e^{-r \log \alpha'(\tau)/2}$$

for an arbitrary real number r. (This definition is possible for complex r. For sake of simplicity we restrict here to automorphic forms of real weight.)

There holds a kind of chain rule for two biholomorphic mappings $\alpha : D \to D'$, $\beta : D' \to D''$:

$$j_r(\beta\alpha, \tau) = w_r(\alpha, \beta) j_r(\beta, \alpha\tau) j_r(\alpha, \tau).$$

Here $w_r(\alpha, \beta)$ is a complex number of absolute value 1. For even r it is one.

Let f be a function on D'. We define the function $f|\alpha$ on D as

$$(f|\alpha)(\tau) = (f\underset{r}{|}\alpha)(\tau) = f(\alpha(\tau)) j_r(\alpha, \tau)^{-1}.$$

Then the chain rule reads as

$$f|(\beta\alpha) = w_r(\alpha, \beta)^{-1} (f|\beta)|\alpha.$$

2.1 Definition. *Let $D \subset \mathbb{C}$ be a domain and let Γ be a group of biholomorphic transformations of D. By a (vector valued) multiplier system of weight $r \in \mathbb{R}$ with values in a finite dimensional complex vector space V we understand a map*

$$v : \Gamma \longrightarrow \mathrm{GL}(V)$$

with the following properties:

1) $v(\gamma_1 \gamma_2) = w_r(\gamma_1, \gamma_2) v(\gamma_1) v(\gamma_2)$.
2) *There exists a positive definite hermitian form on V such that the operators $v(\gamma)$ are unitary.*

Property 1) means that

$$J(\gamma, \tau) = j_r(\gamma, \tau) v(\gamma)$$

is a (vector valued) factor of automorphy, i.e.

$$J(\beta\alpha, \tau) = J(\beta, \alpha\tau) J(\alpha, \tau).$$

So it makes sense to consider functions $f : D \to V$ with the transformation property

$$f(\gamma\tau) = J(\gamma, \tau) f(\tau).$$

2.2 Lemma. *Let $\alpha : D \to \tilde{D}$ be a biholomorphic map of domains and Γ a group of biholomorphic transformations of D. Then $\tilde{\Gamma} = \alpha \Gamma \alpha^{-1}$ is a group of biholomorphic transformations of $\tilde{D}$. Let v be a multiplier system of weight r for (D, Γ), then*

$$\tilde{v}(\gamma) = v(\alpha^{-1} \gamma \alpha) w_r(\alpha^{-1}, \gamma) w_r(\alpha^{-1} \gamma \alpha, \alpha^{-1})$$

is a multiplier system for $(\tilde{D}, \tilde{\Gamma})$ with corresponding automorphy factor

$$\tilde{J}(\gamma, w) = J(\alpha^{-1} \gamma \alpha, \alpha^{-1} w) = \tilde{v}(\gamma) j_r(\gamma, w)^{-1}.$$

Let $f : D \to V$ be a function with the property $f(\gamma\tau) = J(\gamma, \tau) f(\tau)$ for $\gamma \in \Gamma$, then the transformed function $\tilde{f} = f|\alpha^{-1}$ has the transformation property $\tilde{f}(\gamma w) = \tilde{J}(\gamma, w) \tilde{f}(w)$ for $\gamma \in \tilde{\Gamma}$.

§2. Vector valued automorphic forms

From now on Γ denotes a group of biholomorphic transformations of the upper half plane $\mathbb{H}$ such that its inverse image in $\mathrm{SL}(2,\mathbb{R})$ is discrete. We denote by $S \subset \mathbb{R} \cup \{\infty\}$ the set of cusps of this group and we denote by $\mathbb{H}^* = \mathbb{H} \cup S$ the extended upper half plane. We assume that the Riemann surface

$$X = X_\Gamma := \mathbb{H}^*/\Gamma$$

is compact. (It can be shown that this is equivalent to the fact that $\mathbb{H}/\Gamma$ has finite volume, see Lemma 9.1 for one direction.) Then the set of cusp classes is finite.

In the following we fix a group Γ, a real number r, and a multiplier system $v : \Gamma \to \mathrm{GL}(V)$ of weight r for Γ. Let n be the dimension of V. We want to define for each point $a \in \mathbb{H}^*$ an unordered n-tuple of numbers

$$\xi_1, \ldots, \xi_n, \quad 0 \le \xi_i < 1 \quad (n = \dim V).$$

We will call them the *characteristic numbers* of a (with respect to Γ, r, v).

We start with the case where $a \in \mathbb{H}$ is an inner point. We transform it to the origin of the unit disk $\mathbb{E}$ by means of the transformation

$$\alpha(\tau) = \frac{\tau - a}{\tau - \bar{a}}.$$

We consider the conjugate group $\tilde{\Gamma} = \alpha \Gamma \alpha^{-1}$. We also consider the conjugate multiplier system $\tilde{v}$ and corresponding automorphy factor $\tilde{J}$ in the sense of Lemma 2.2.

2.3 Remark and Definition. *For a point $a \in \mathbb{H}$ we consider the conjugate group*

$$\tilde{\Gamma} = \alpha \Gamma \alpha^{-1}, \quad \alpha(\tau) = \frac{\tau - a}{\tau - \bar{a}},$$

and the transformed automorphy factor $\tilde{J}(\gamma, w)$. The stabilizer of the origin in $\tilde{\Gamma}$ is generated by the transformation $r_e(w) = e^{2\pi i/e} w$ where e is the order of the stabilizer Γ_a. The transformation $R = \tilde{J}(r_e, w)$ is independent of w and has the property $R^e = \mathrm{id}$. We define the characteristic numbers

$$\xi_1, \ldots, \xi_n, \quad 0 \le \xi_i < 1,$$

such that $e^{2\pi i \xi_\nu}$ are the eigenvalues of R. The numbers $e\xi_\nu$ are integral.

The proof is rather trivial. Every element $\gamma \in \tilde{\Gamma}$ which stabilizes the origin must be of the form $w \mapsto \zeta w$ where ζ is a complex number of absolute number one. The only subgroup of order e of the multiplicative group of complex numbers is the group generated by $e^{2\pi i/e}$. Since the derivative of γ is constant, $\tilde{J}(\gamma, w)$ is independent of w. The automorphy property implies that it is a homomorphism. The image is a group of some order that divides e. □

Another way to describe the characteristic numbers is as follows.

2.4 Remark. *(Notations as in Remark and Definition 2.3.) Let β be any biholomorphic map from $\mathbb{H}$ to $\mathbb{E}$ with the property $\beta(a) = 0$. Consider in the stabilizer Γ_a the generator γ that corresponds to the element $r_e(w) = e^{2\pi i/e}w$ in the group $\beta\Gamma_a\beta^{-1}$. Then the characteristic numbers ξ_ν are defined such that $0 \leq \xi_\nu < 1$ and such that $e^{2\pi i\xi_\nu}$ are the eigenvalues of $J(\gamma, a)$.*

Supplement. *The characteristic numbers depend only on the Γ-equivalence class of a.*

Proof. In the case where β is the transformation α this follows from the formula $\tilde{J}(\gamma, w) = J(\alpha^{-1}\gamma\alpha, \alpha^{-1}w)$ (Lemma 2.2). In the general case one has to use that a general β and α are related by a rotation $\beta(\tau) = \zeta\alpha(\tau)$.

For the proof of the supplement we consider an equivalent point $b = \gamma(a)$. We transform a to 0 by α and then use $\beta = \gamma(a)$ to transform b to zero. For the two corresponding generators $\gamma_a \in \Gamma_a$ and $\gamma_b \in \Gamma_b$ we then have $\gamma_b = \gamma\gamma_a\gamma^{-1}$. The chain rules for $\gamma_b\gamma$ and $\gamma\gamma_a$ show that $J(\gamma_a, a)$ and $J(\gamma_b, b)$ are conjugate. □

Next we define the characteristic numbers in the case where a is the cusp ∞. The stabilizer Γ_∞ is generated by a translation

$$t^N(\tau) := \tau + N, \quad N > 0.$$

The matrix

$$R = J(t^N, \tau)$$

is independent of τ. Its eigenvalues have absolute value 1. The characteristic numbers are defined such that $e^{2\pi i\xi_\nu}$ are the eigenvalues of R.

We treat the case where a is an arbitrary cusp. We choose a transformation $\alpha \in \mathrm{Aut}(\mathbb{H})$ with the property $\alpha(a) = \infty$. We can consider the conjugate group $\tilde{\Gamma} = \alpha\Gamma\alpha^{-1}$ and the conjugate multiplier system $\tilde{v}$ (and of course the same r). The group $\tilde{\Gamma}$ has the cusp ∞. We want to define the characteristic numbers of (Γ, r, v) at the cusp a to be the characteristic numbers of $(\tilde{\Gamma}, r, \tilde{v})$ at ∞. It is easy to prove that this definition does not depend on the choice of α.

2.5 Lemma and Definition. *Let ∞ be a cusp of Γ and let t^N be the translation in Γ with smallest positive N. The characteristic numbers*

$$\xi_1 \ldots, \xi_n, \quad 0 \leq \xi_i < 1,$$

are defined such that $e^{2\pi i\xi_\nu}$ are the eigenvalues of $R = J(t^N, \tau)$.

Let a be an arbitrary cusp and let $\alpha \in \mathrm{Aut}(\mathbb{H})$ be a transformation with the property $\alpha(a) = \infty$. We consider the conjugate group $\tilde{\Gamma} = \alpha\Gamma\alpha^{-1}$ and the conjugate multiplier system $\tilde{v}$. The characteristic numbers of $(\tilde{\Gamma}, r, \tilde{v})$ at ∞ are independent of the choice of α. They are called the characteristic numbers of the cusp a.

§2. Vector valued automorphic forms

Proof of the second part. We can assume that $a = \infty$. Then α is of the form $\alpha(\tau) = u\tau + v$. Let $t^N(\tau) = \tau + N$ be the generator of Γ_∞. We set $\tilde{N} = uN$. Then $t^{\tilde{N}} = \alpha t^N \alpha^{-1}$ and this is the generator of $\tilde{\Gamma}_\infty$. Taking a suitable basis of V we can assume that the matrix of $R = J(t^N, \cdot)$ is diagonal. Then we can assume that V has dimension one and that R acts by multiplication by $e^{2\pi i x}$ where x is the characteristic number. The function $f(\tau) = e^{2\pi i x \tau / N}$ then has the property $f(\tau + N) = J(t^N, \tau) f(\tau)$. From the second part of Lemma 2.2 follows that the function $\tilde{f}(\tau) = f(u\tau + v)$ has the property

$$\tilde{f}(\tau + \tilde{N}) = \tilde{J}(t^{\tilde{N}}, \tau)\tilde{f}(\tau).$$

This formula implies that $\tilde{J}(t^{\tilde{N}}, \tau)$ is also the multiplication by $e^{2\pi i x}$. □

2.6 Lemma. *The characteristic numbers depend only on the Γ-orbit of a point $a \in \mathbb{H}^*$. Hence they can be considered for points $x \in X_\Gamma$. They can be different from $(0, \ldots, 0)$ only for cusps or elliptic fixed points.*

We want to introduce *local automorphic forms*. Let $U \subset X_\Gamma$ be an open subset. We denote its inverse image in $\mathbb{H}^*$ by $\tilde{U}$. This is a Γ-invariant subset. Hence we can consider all holomorphic functions $f : \tilde{U} - S \to \mathbb{C}$ with the transformation property

$$f(\gamma\tau) = J(\gamma, \tau)f(\tau), \qquad \gamma \in \Gamma.$$

Assume that Γ has cusp ∞ and that it is contained in $\tilde{U}$. Then $\tilde{U}$ contains some upper half plane $\operatorname{Im} \tau > C > 0$ and f has the transformation property $f(\tau + N) = Rf(\tau)$. Since R has finite order, f has some multiple of N as period. We call f regular at ∞ if f is bounded for $\operatorname{Im} \tau \to \infty$ and cuspidal if it tends to 0. We can diagonalize R and describe f by components

$$f_\nu(\tau + N) = e^{2\pi i \xi_\nu} f_\nu(\tau).$$

The function $g_\nu(\tau) = f_\nu(\tau)e^{-2\pi i \xi_\nu \tau / N}$ has period N. Hence we have a Fourier expansion

$$f_\nu(\tau) = e^{\frac{2\pi i}{N}\xi_\nu \tau} \sum_{m=-\infty}^{\infty} b_\nu(m) e^{\frac{2\pi i}{N} m \tau}.$$

We want to simplify the notations and introduce for this the symbolic notation

$$q^a := e^{2\pi i a \tau} \qquad (a \in \mathbb{C}).$$

Then we can write the expansion in the simplified form

$$\boxed{f_\nu(\tau) = \sum_{m \in \Lambda_\nu} a_\nu(m) q^m \quad \text{where} \quad \Lambda_\nu = \frac{1}{N}(\xi_\nu + \mathbb{Z})}$$

The function f is called regular at ∞ if $a_\nu(m) \ne 0$ implies $m \ge 0$ and cuspidal if it implies $m > 0$.

Using "transformation to ∞" one can define the notions "regular" and "cuspidal" also for other cusps. It is clear that this notion does not depend on the choice of the transformation. It is also clear that this notion depends only on the Γ-orbit of a cusp.

We define a certain sheaf $\mathcal{M} = \mathcal{M}_\Gamma(r, v)$ on X_Γ. For open U in $X = X_\Gamma$ the space $\mathcal{M}(U)$ consists of all local automorphic forms $f : \tilde{U} - S \to V$ which are regular at the cusps. This defines a sheaf and even more an $\mathcal{O}_X$-module. For even r and the trivial one-dimensional representation v we write $\mathcal{M}(r)$ instead of $\mathcal{M}(r, v)$. We also can consider the subsheaf $\mathcal{M}^{\text{cusp}} = \mathcal{M}_\Gamma^{\text{cusp}}(r, v)$ of all cuspidal local automorphic forms. This is also an $\mathcal{O}_X$-module.

2.7 Lemma. *The sheafs $\mathcal{M} = \mathcal{M}_\Gamma(r, v)$, $\mathcal{M}^{\text{cusp}} = \mathcal{M}_\Gamma^{\text{cusp}}(r, v)$ are vector bundles of rank $n = \dim V$, hence coherent.*

Proof. We restrict to $\mathcal{M}$ since the case $\mathcal{M}^{\text{cusp}}$ is similar. We want to show that $\mathcal{M}_x$ is a free $\mathcal{O}_{X,x}$-module for each $x \in X_\Gamma$.

First we assume that a is a cusp. We can restrict to $a = \infty$. As explained above, the elements of $\mathcal{M}_x$ can be considered as Fourier series of the kind

$$f_\nu(\tau) = \sum_{m \in \mathbb{Z},\ m + \xi_\nu \ge 0} b_\nu(m) e^{\frac{2\pi i}{N}(m + \xi_\nu)\tau}.$$

Recall that $0 \le \xi_\nu < 1$. Hence $m + \xi_\nu \ge 0$ means the same as $m \ge 0$. If we map $f_\nu(\tau)$ to

$$\sum_{m \ge 0} b_\nu(m) e^{\frac{2\pi i}{N} m\tau},$$

we get an isomorphism from $\mathcal{M}_x$ to $\mathcal{O}_{X,x}^n$. This shows that $\mathcal{M}_x$ is free.

Next we consider the case where a is an innner point. In this case we can identify $\mathcal{M}_x$ with holomorphic functions f in a small disk around $w = 0$ which transform as

$$f(e^{2\pi i/e} w) = R f(w), \quad R^e = \text{id}.$$

The components of f with respect to a basis of eigenvectors satisfy

$$f_\nu(e^{2\pi i/e} w) = e^{2\pi i \xi_\nu} f_\nu(w).$$

From the Taylor expansion one can derive that $f_\nu(w) = w^{e\xi_\nu} g_\nu(w^e)$. The local ring $\mathcal{O}_{X,x}$ can be identified with the ring of power series $\mathbb{C}\{w^e\}$. The map $f_\nu \mapsto g_\nu$ gives an $\mathcal{O}_{X,x}$-linear isomorphism from $\mathcal{M}_x$ to $\mathcal{O}_{X,x}^n$.

So we have shown that $\mathcal{M}_x$ is free for all x. The statement that $\mathcal{M}$ is a vector bundle means a little more, namely that a basis of $\mathcal{M}_x$ gives a basis in a full neighborhood of x. This follows from the following trivial description of

§2. Vector valued automorphic forms

$\mathcal{M}_x$ in the case where $a \in \mathbb{H}$ is not a fixed point: let $f_1, \ldots, f_n$ be holomorphic functions in an open neighborhood of a which do not vanish at a. Denote by $e_1, \ldots, e_n$ the unit vectors. Then $f_1 e_1, \ldots, f_n e_n$ defines a basis of the module $\mathcal{M}_x$.

This completes the proof of Lemma 2.7. □

For later purpose we collect the local description of the sheaf $\mathcal{M}(r, v)$ that we obtained during the proof of Lemma 2.7.

2.8 Lemma. *Let ∞ be a cusp of Γ. Then we have*

$$\mathcal{M}(r,v)_{[\infty]} \cong \bigoplus e^{2\pi i \xi_\nu \tau / N} \mathcal{O}_{X_\Gamma, [\infty]} e_\nu.$$

Let $a \in \mathbb{H}$ be an interior point. Then we have

$$\mathcal{M}(r,v)_{[a]} \cong \bigoplus w^{e\xi_\nu} \mathcal{O}_{X_\Gamma, [a]} e_\nu.$$

For a multiplier system v of weight r with values in the vector space V one can define the dual multiplier system v'. It is realized on $\mathrm{Hom}_{\mathbb{C}}(V, \mathbb{C})$. By definition $v'(\gamma)$ is the transposed of $v(\gamma^{-1})$. It is easy to check that this is a multiplier system of weight $-r$. We mention that a multiplier system of weight r can be considered as multiplier system of weight r' for each $r' \equiv r$ mod 2.

As in the case of the sheaf $\mathcal{M}$, we write $\mathcal{M}^{\mathrm{cusp}}(r)$ instead of $\mathcal{M}^{\mathrm{cusp}}(r, v)$ for even r and the trivial one-dimensional multiplier system v.

2.9 Lemma. *The sheaf $\mathcal{M}^{\mathrm{cusp}}(2)$ is a canonical sheaf. The dual sheaf of $\mathcal{M}(r, v)$ is isomorphic to $\mathcal{M}^{\mathrm{cusp}}(2-r, v')$, where v' denotes the dual multiplier system.*

Proof. The canonical sheaf on a compact Riemann surface is the sheaf of holomorphic differentials. Let ω be a holomorphic differential on an open subset $U \subset X_\Gamma$. Its inverse image on $\tilde{U} - S$ is of the form $f(\tau)d\tau$. The function f transforms like an automorphic form of weight two (and trivial multiplier system). Using the formula

$$2\pi i d\tau = dq/q \quad \text{for} \quad q = e^{2\pi i \tau}$$

it is easy to show that the regularity of ω at the cusp classes means that f is cuspidal. For the elliptic fixed points a similar argument works. We omit it. □

Next we define a pairing

$$\mathcal{M}(r,v) \times \mathcal{M}^{\mathrm{cusp}}(2-r, v') \longrightarrow \mathcal{M}^{\mathrm{cusp}}(2).$$

For this we use the natural pairing

$$V \times \mathrm{Hom}(V, \mathbb{C}) \longrightarrow \mathbb{C}, \quad \langle v, L \rangle = L(v).$$

Let $f \in \mathcal{M}(r, v)(U)$ and $g \in \mathcal{M}^{\mathrm{cusp}}(2 - r, v')(U)$ be local automorphic forms on some $U \subset X_\Gamma$. Then $\langle f, g \rangle$ transforms like an automorphic form of weight 2 with respect to the trivial multiplier system. It is clear that it is cuspidal. So the pairing has been defined. It has to be checked that it is non-degenerated. This can be done by a local computation at points $x \in X_\Gamma$. We restrict to the case where x is the image of the cusp ∞. Recall that – using a suitable basis of V – the elements of $\mathcal{M}(r, v)_x$ can be identified with Fourier series

$$f_\nu(\tau) = \sum_{m+\xi_\nu \geq 0}^{\infty} a_\nu(m) e^{\frac{2\pi \mathrm{i}}{N}(m+\xi_\nu)\tau}.$$

The characteristic numbers y_ν of the dual multiplier system have the property $\xi_\nu + y_\nu \equiv 0 \bmod 1$. Hence – using the dual basis – the elements of $\mathcal{M}^{\mathrm{cusp}}(2 - r, v')_x$ can be identified with Fourier series

$$g_\nu(\tau) = \sum_{m-\xi_\nu > 0}^{\infty} b_\nu(m) e^{\frac{2\pi \mathrm{i}}{N}(m-\xi_\nu)\tau}$$

and the pairing is just $\sum f_\nu g_\nu$. The condition $m + \xi_\nu \geq 0$ is equivalent to $m \geq 0$ and the condition $m - \xi_\nu > 0$ is equivalent to $m \geq 1$. Finally $\mathcal{M}^{\mathrm{cusp}}(2)_x$ can be identified with all Fourier series

$$h(\tau) = \sum_{m \geq 1} c(m) e^{\frac{2\pi \mathrm{i}}{N} m\tau}.$$

Let $q = e^{\frac{2\pi \mathrm{i}}{N}\tau}$. Using the isomorphisms

$$\mathcal{M}(r, v)_x \xrightarrow{\sim} \mathbb{C}\{q\}^n, \quad f \longmapsto \left(\sum a_\nu(m) q^m\right),$$
$$\mathcal{M}^{\mathrm{cusp}}(2 - r, v')_x \xrightarrow{\sim} \mathbb{C}\{q\}^n, \quad f \longmapsto \left(\sum b_\nu(m) q^{m-1}\right),$$
$$\mathcal{M}(2)_x \xrightarrow{\sim} \mathbb{C}\{q\}^n, \quad h \longmapsto \left(\sum c_\nu(m) q^{m-1}\right),$$

the pairing gets equivalent to the standard pairing

$$\mathbb{C}\{q\}^n \times \mathbb{C}\{q\}^n \longrightarrow \mathbb{C}\{q\}, \quad \langle P, Q \rangle = \sum_\nu P_\nu Q_\mu,$$

which is obviously non-degenerated. $\square$

3. Direct and inverse images

Let $f : X \to Y$ be a continuous map of topological spaces and let F be a presheaf of abelian groups on X. Then one can define for an arbitrary open subset $V \subset Y$
$$(f_*F)(V) := F(f^{-1}(V)).$$
With obvious restriction maps this defines a presheaf f_*F on X. It is called the *direct image*. If F is a sheaf, then f_*F is a sheaf too. A homomorphism of presheaves $F \to G$ induces a natural homomorphism $f_*F \to f_*G$. We also mention that there is a natural homomorphism
$$(f_*F)_{f(a)} \longrightarrow F_a \qquad (a \in X).$$
There is a similar, but not quite so easy construction, which associates to a presheaf G on Y a presheaf, actually a sheaf, $f^{-1}G$ on X. In the case of the identity map we will obtain the generated sheaf $\hat{F}$. If $i : U \hookrightarrow X$ is the canonical inclusion of an open subset and if G is a sheaf on X, then $i^{-1}G$ is naturally isomorphic to the restriction $G|U$. Actually, we will construct a subgroup
$$(f^{-1}G)(U) \subset \prod_{a \in U} G_{f(a)}.$$
By definition, a family $(t_a)_{a \in U}$, $t_a \in G_{f(a)}$, belongs to this subgroup if it is compatible in the following obvious sense. For each $a \in U$ there exists a small open neighborhood $f(a) \in V \subset Y$ and a section $t \in G(V)$ with the property $t_a = [V,t]_{f(a)}$ for all $a \in U$ such that $f(a) \in V$. It is easy to verify that this defines a sheaf $f^{-1}G$. A homomorphism of presheaves $G \to H$ induces a natural homomorphism $f^{-1}G \to f^{-1}H$. Notice that $\mathrm{id}^{-1}G$ equals the generated sheaf. For an open neighborhood $a \in U \subset X$ there is a natural projection homomorphism $(f^{-1}G)(U) \longrightarrow G_{f(a)}$. It induces a homomorphism $(f^{-1}G)_a \to G_{f(a)}$. This is actually an isomorphism.

3.1 Lemma. *Let $f : X \to Y$ be a continuous map and let G be a sheaf on Y. There is a natural isomorphism*
$$(f^{-1}G)_a \xrightarrow{\sim} G_{f(a)}.$$

Let $V \subset Y$ be an open subset. For each $a \in f^{-1}(V)$ we can consider the natural homomorphism $G(V) \to G_{f(a)}$ and collect them to
$$G(V) \longrightarrow \prod_{a \in f^{-1}(V)} G_{f(a)}.$$
The right hand side contains $f^{-1}(G)(f^{-1}(V)) = f_*f^{-1}(G)(V)$. It is easy to check that the image of $G(V)$ is contained in this subgroup. So we obtain a homomorphism $G(V) \to f_*f^{-1}(G)(V)$. It is easy to check that this gives a homomorphism of sheaves.

3.2 Lemma. *Let $f : X \to Y$ be a continuous map and let G be a sheaf on Y. There is a natural homomorphism of sheaves*

$$G \longrightarrow f_* f^{-1}(G).$$

Now we consider a sheaf F on X and we consider the sheaf $f^{-1} f_* F$. Let $U \subset X$ be an open subset. Then $(f^{-1} f_* F)(U)$ is contained in $\prod_{a \in U} (f_* F)_{f(a)}$ which can be identified with $\prod_{a \in U} F_a$. The module $F(U)$ is embedded in this product. It is easy to check that we obtain a homomorphism of $f^{-1} f_* F$ into F.

3.3 Lemma. *Let $f : X \to Y$ be a continuous map and let F be a sheaf on X. There is a natural homomorphism of sheaves*

$$f^{-1} f_* F \longrightarrow F.$$

Assume now that a sheaf F on X and a sheaf G on Y is given. Let $f^{-1} G \to F$ be a homomorphism. It induces a homomorphism $f_* f^{-1} G \to f_* F$, and, making use of Lemma 3.2, we get a homomorphism $G \to f_* F$. Conversely, let $G \to f_* F$ be a homomorphism. It induces $f^{-1} G \to f^{-1} f_* F$ and, by means of Lemma 3.3, we get a homomorphism $f^{-1} G \to F$. It is easy to check that the two construction are inverse. If we denote by $\operatorname{Hom}(F_1, F_2)$ the set of all homomorphisms of one sheaf into another (on the same space), then we can formalize this as follows.

3.4 Proposition. *Let $f : X \to Y$ be a continuous map, let F be a sheaf on X, and let G be a sheaf on Y. There is a natural bijection*

$$\operatorname{Hom}(f^{-1} G, F) \xrightarrow{\sim} \operatorname{Hom}(G, f_* F).$$

If one specializes this formula to $G = f_* F$, then one can consider the identity on the right hand side. The corresponding homomorphism on the left-hand side is that in Lemma 3.3. If one specializes it to $F = f^{-1} G$, the identity on the left hand side corresponds to Lemma 3.2.

Direct and inverse images of modules

We now assume that we have a morphism $f : (X, \mathcal{O}_X) \to (Y, \mathcal{O}_Y)$ of geometric spaces. Recall that we have for each open subset $V \subset Y$ a natural homomorphism $\mathcal{O}_Y(V) \to \mathcal{O}_X(f^{-1} V)$. This can be read as a homomorphism of sheaves $\mathcal{O}_Y \to f_* \mathcal{O}_X$. In fact, this is a homomorphism of sheaves of rings. (This means that the occurring homomorphisms are homomorphisms of rings

§3. Direct and inverse images

and not only of abelian groups.) Using Proposition 3.4 we obtain a homomorphism $f^{-1}\mathcal{O}_Y \to \mathcal{O}_X$. It is easy to verify that $f^{-1}\mathcal{O}_Y$ is a sheaf of rings and that this homomorphism is a homomorphism of sheaves of rings. In particular, $\mathcal{O}_X$ carries a natural structure as $f^{-1}\mathcal{O}_Y$-module. Assume that $\mathcal{M}$ is an $\mathcal{O}_X$-module. Then $f_*\mathcal{M}$ carries a natural structure as $f_*\mathcal{O}_X$-module. Using the homomorphism $\mathcal{O}_Y \to f_*\mathcal{O}_X$, we obtain a structure as $\mathcal{O}_Y$-module. We say simply that $f_*\mathcal{M}$ is an $\mathcal{O}_Y$-module if $\mathcal{M}$ is an $\mathcal{O}_X$-module.

The situation for the inverse image is slightly more complicated. Let $\mathcal{N}$ be an $\mathcal{O}_Y$-module. It is no problem to equip $f^{-1}\mathcal{N}$ with a structure as $f^{-1}\mathcal{O}_Y$-module but there is no natural way to get a structure as $\mathcal{O}_X$-module. What we can do is to consider

$$f^*\mathcal{N} := f^{-1}\mathcal{N} \otimes_{f^{-1}\mathcal{O}_Y} \mathcal{O}_X.$$

This carries a natural structure as $\mathcal{O}_X$-module. An $\mathcal{O}_Y$-linear map $\mathcal{N}_1 \to \mathcal{N}_2$ induces an $\mathcal{O}_X$-linear map $f^*\mathcal{N}_1 \to f^*\mathcal{N}_2$. If we denote the set of all $\mathcal{O}_Y$-linear maps between $\mathcal{N}_1$ and $\mathcal{N}_2$ by $\mathrm{Hom}_{\mathcal{O}_Y}(\mathcal{N}_1,\mathcal{N}_2)$, we obtain the following analogue of Proposition 3.4.

3.5 Proposition. *Let $f : (X,\mathcal{O}_X) \to (Y,\mathcal{O}_Y)$ be a morphism of geometric spaces, let $\mathcal{M}$ be an $\mathcal{O}_X$-module, and let be $\mathcal{N}$ be a $\mathcal{O}_Y$-module. There is a natural bijection*

$$\mathrm{Hom}_{\mathcal{O}_X}(f^*\mathcal{N},\mathcal{M}) \xrightarrow{\sim} \mathrm{Hom}_{\mathcal{O}_Y}(\mathcal{N},f_*\mathcal{M}).$$

This includes natural homomorphisms

$$\mathcal{N} \longrightarrow f_*f^*\mathcal{N}, \quad f^*f_*\mathcal{M} \longrightarrow \mathcal{M}.$$

3.6 Proposition. *Let $f : (X,\mathcal{O}_X) \to (Y,\mathcal{O}_Y)$ be a morphism of geometric spaces. There is a natural isomorphism*

$$f^*\mathcal{O}_Y \cong \mathcal{O}_X.$$

Let $\mathcal{N}$ be a vector bundle. Then $f^*\mathcal{N}$ is a vector bundle too. The rank is preserved.

Proof. By definition,

$$f^*\mathcal{O}_Y = f^{-1}\mathcal{O}_Y \otimes_{f^{-1}\mathcal{O}_Y} \mathcal{O}_X \cong \mathcal{O}_X.$$

An easy consequence is $f^*\mathcal{O}_Y^n \cong \mathcal{O}_X^n$. The constructions f^{-1}, the tensor product and f^* are all compatible with restriction to open subsets, for example $(f^*\mathcal{N})|U \cong f^*(\mathcal{N}|U)$. This proves Proposition 3.6. □

There is an easy formula for the stalk of the inverse image. Actually this formula is true for vector bundles but not for arbitrary $\mathcal{O}_Y$-modules.

3.7 Lemma. *Let $f : (X, \mathcal{O}_X) \to (Y, \mathcal{O}_Y)$ be a morphism of geometric spaces, let $\mathcal{N}$ be a vector bundle on Y, and let a be a point in X. There is a natural isomorphism*

$$(f^*\mathcal{N})_a \cong \mathcal{N}_{f(a)} \otimes_{\mathcal{O}_{Y,f(a)}} \mathcal{O}_{X,a}.$$

Proof. It is sufficient to prove this for $\mathcal{N} = \mathcal{O}_Y$. In this case the proof is trivial. □

Inverse image of divisors

The inverse image of a line bundle $\mathcal{L}$ is a line bundle too. Now we assume that the geometric spaces X, Y are Riemann surfaces and that $f : X \to Y$ is a holomorphic map which is not constant on any connected component of X. Recall that we defined for each $a \in X$ the multiplicity of f at a (Definition V.5.7). We denote it by $\mathrm{Ord}(f, a)$. Assume that $\mathcal{L} = \mathcal{O}_D$ is the line bundle associated to a divisor. We will show that $f^*\mathcal{L}$ is related to the following divisor.

$$(f^*D)(a) := \mathrm{Ord}(f, a) D(f(a)).$$

3.8 Lemma. *Let $f : X \to Y$ be a holomorphic map of Riemann surfaces which is not constant on any connected component of X and let D be a divisor on Y. Then there exists a natural isomorphism*

$$f^*\mathcal{O}_D \cong \mathcal{O}_{f^*D}.$$

Proof. Let $V \subset Y$ be an open subset. We obtain a map

$$\mathcal{O}_D(V) \longrightarrow \mathcal{O}_{f^*D}(f^{-1}(V)), \quad g \longmapsto g \circ f.$$

We can read it as a map of sheaves $\mathcal{O}_D \to f_*\mathcal{O}_{f^*D}$. The universal property Proposition 3.5 gives an $\mathcal{O}_X$-linear map $f^*\mathcal{O}_D \to \mathcal{O}_{f^*D}$. We have to show that this is an isomorphism. Since this is a local question, we can assume that $D = (g)$ is the divisor of a meromorphic function. Multiplication by g and by $g \circ f$ gives trivializations $\mathcal{O}_Y \cong \mathcal{O}_D$, $\mathcal{O}_X \cong \mathcal{O}_{f^*D}$. This gives a reduction to the zero divisor. But then the statement follows from Proposition 3.6. □

3.9 Proposition. *Let $f : X \to Y$ be a non-constant holomorphic map of connected Riemann surfaces and let $\mathcal{N}_1 \to \mathcal{N}_2 \to \mathcal{N}_3$ be an exact sequence of vector bundles on Y. Then the sequence*

$$f^*\mathcal{N}_1 \longrightarrow f^*\mathcal{N}_2 \longrightarrow f^*\mathcal{N}_3$$

is exact too.

Proof. Looking at the stalks, it is clear that $f^{-1}\mathcal{N}_1 \longrightarrow f^{-1}\mathcal{N}_2 \longrightarrow f^{-1}\mathcal{N}_3$ is exact. It remains to show that the sequence remains exact after tensoring it with $\mathcal{O}_{X,a}$ over $\mathcal{O}_{Y,f(a)}$. Actually, $\mathcal{O}_{X,a}$ is a *free* $\mathcal{O}_{Y,f(a)}$-module. This shows the local behaviour of f. We can assume that f is the map $z \mapsto z^n$ from the unit disc into itself. But then $\mathcal{O}_{X,a}$ can be identified with the ring $\mathcal{O}\{z\}$ of convergent power series and $\mathcal{O}_{Y,f(a)}$ with the subring $\mathbb{C}\{z^n\}$. We get a free module with basis $1, z, \ldots, z^{n-1}$. □

3.10 Proposition. *Let $f : X \to Y$ be a non-constant holomorphic map of connected and compact Riemann surfaces and let $\mathcal{N}$ be a vector bundle on Y. Then one has*
$$\deg f^*\mathcal{N} = \deg(f) \deg \mathcal{N}.$$

Recall that $\deg(f)$ is the covering degree, i.e. the number of inverse points of a single point in Y, counted with multiplicity.

Proof. The statement is clear if $\mathcal{N}$ is a line bundle that comes from a divisor. But then it is true for any line bundle. The general case can be settled by induction on the rank, making use of Lemma VI.3.6 and Proposition 3.9. □

4. The computation of the degree, first method

The first method to compute the degree needs some more assumptions about (r, v), but it has the advantage to be very simple. In particular, the volume of the fundamental domain and the Gauss–Bonnet formula will not be used. For practical use, as in the theory of Borcherds products, this special case is good enough. Hence we will describe this method in some detail. In the Appendix we will sketch, how these assumptions can be avoided.

4.1 Assumption. *The triple Γ, r, v has the following property.*
1) *The weight r is rational.*
2) *The matrices $v(\gamma)$ are of finite order.*
3) *There exists a subgroup $\Gamma_0 \subset \Gamma$ of finite index such that $v(\gamma)$ can be simultaneously diagonalized for $\gamma \in \Gamma_0$.*
4) *There exists a normal subgroup of finite index $\Gamma_0 \subset \Gamma$ such that Γ_0 acts fixed point free on $\mathbb{H}$ and such that the characteristic numbers of all cusps with respect to Γ_0 are zero.*

For the rest of this section we take this assumption to be granted. We restrict v to Γ_0 and consider the sheaf

$$\mathcal{M}_0 = \mathcal{M}_{\Gamma_0}(r, v)$$

on the Riemann surface X_{Γ_0}. We want to compare the degrees of $\mathcal{M}$ and $\mathcal{M}_0$. Let
$$\pi : X_{\Gamma_0} \longrightarrow X_\Gamma$$
be the natural covering. There is an obvious inclusion of sheaves $\mathcal{M} \hookrightarrow \pi_* \mathcal{M}_0$. By functoriality this induces a map
$$\pi^* \mathcal{M} \longrightarrow \mathcal{M}_0.$$
Let $x \in X_{\Gamma_0}$. The stalk of $\pi^* \mathcal{M}$ is
$$(\pi^* \mathcal{M})_a \cong \mathcal{M}_{\pi(x)} \otimes_{\mathcal{O}_{X_{\Gamma,\pi(x)}}} \mathcal{O}_{X_{\Gamma_0},x}.$$
Since $\mathcal{O}_{X_{\Gamma_0},x}$ is a free $\mathcal{O}_{X_{\Gamma,\pi(x)}}$-module, we see that
$$(\pi^* \mathcal{M})_{\pi(x)} \longrightarrow (\mathcal{M}_0)_x$$
is injective. Outside a finite set (images of cusps and of elliptic fixed points of Γ) it is an isomorphism. So we get an exact sequence
$$0 \longrightarrow \pi^* \mathcal{M} \longrightarrow \mathcal{M}_0 \longrightarrow \mathcal{K} \longrightarrow 0$$
with a skyscraper sheaf $\mathcal{K}$. We have to compute its degree
$$\deg \mathcal{K} = \sum_{x \in X_{\Gamma_0}} \dim \mathcal{K}_x.$$
We compute
$$\mathcal{K}_x = (\mathcal{M}_0)_x \,/\, (\mathcal{M}_{\pi(x)} \otimes_{\mathcal{O}_{X_{\Gamma,\pi(x)}}} \mathcal{O}_{X_{\Gamma_0},x})$$
first in the case where x comes from an inner point $a \in \mathbb{H}$. Let $w = (\tau - a)/(\tau - \bar{a})$. Recall that we assume that a is not an elliptic fixed point of Γ_0. Then the local ring $\mathcal{O}_{X_{\Gamma_0},x}$ can be identified with the ring of power series $\mathbb{C}\{w\}$. The ring $\mathcal{O}_{X_\Gamma,\pi(x)}$ can be identified with $\mathbb{C}\{w^e\}$. As in section two we take a basis of V such that all $v(\gamma)$ are diagonal. Then we have natural isomorphisms
$$(\mathcal{M}_0)_x \cong \mathbb{C}\{w\}^n.$$
and
$$\mathcal{M}_{\pi(x)} = \prod_{\nu=1}^n w^{e\xi_\nu} \mathbb{C}\{w^e\}.$$
If we tensor this with $\mathbb{C}\{w\}$ we get
$$\mathcal{M}_{\pi(x)} \otimes_{\mathcal{O}_{X_\Gamma,\pi(x)}} \mathcal{O}_{X_{\Gamma_0},x} = \prod_{\nu=1}^n w^{e\xi_\nu} \mathbb{C}\{w\}.$$
This shows the following result.

§4. The computation of the degree, first method

4.2 Lemma. Let $\pi : X_{\Gamma_0} \to X_\Gamma$ be the natural projection. We consider a point $x \in X_{\Gamma_0}$ which is the image of an inner point $a \in \mathbb{H}$. Let v be a multiplier system of weight r for Γ. We denote by
$$\sigma(a) = \xi_1 + \cdots + \xi_n$$
the sum of the characteristic numbers at a. Then the formula
$$\dim \mathcal{K}_x = e\sigma(a) \qquad (\mathcal{K} = \mathcal{M}_{\Gamma_0}(r,v)/\pi^*\mathcal{M}_\Gamma(r,v))$$
holds.

Now we consider the case that $x \in X_{\Gamma_0}$ is the image of the cusp ∞. Analogously to N for Γ, we denote by N_0 the smallest positive number such that $\tau \mapsto \tau + N_0$ is in Γ_0. The number N_0/N is integral. We set
$$q = e^{\frac{2\pi i}{N_0}\tau}.$$
Then the local ring of X_{Γ_0} at x is $\mathbb{C}\{q\}$ and the local ring of X_Γ at $\pi(x)$ is $\mathbb{C}\{q^{N_0/N}\}$. The stalk of $\mathcal{M} = \mathcal{M}_\Gamma(r,v)$ at $\pi(x)$ is (after diagonalization)
$$\mathcal{M}_{\pi(x)} = \prod_{\nu=1}^{n} e^{\frac{2\pi i}{N}\xi_\nu \tau}\mathbb{C}\{q^{N_0/N}\}.$$
We get
$$\mathcal{M}_{\pi(x)} \otimes_{\mathcal{O}_{\Gamma,\pi(x)}} \mathcal{O}_{X_{\Gamma_0},x} = \prod_{\nu=1}^{n} e^{\frac{2\pi i}{N}\xi_\nu \tau}\mathbb{C}\{q\}.$$
The characteristic numbers $y_1, \ldots, y_n$ with respect to Γ_0 are defined by
$$y_\nu \equiv (N_0/N)\xi_\nu \text{ ord } 1, \qquad 0 \le y_\nu < 1,$$
or, using the Gauss bracket,
$$y_\nu = (N_0/N)\xi_\nu - [(N_0/N)\xi_\nu].$$
Hence the stalk of $\mathcal{M}_0 = \mathcal{M}_{\Gamma_0}(r,v)$ at x is
$$\mathcal{M}_{0,x} = \prod_{\nu=1}^{n} e^{\frac{2\pi i}{N_0}y_\nu \tau}\mathbb{C}\{q\}.$$
This shows
$$\dim \mathcal{K}_x = \sum_{\nu=1}^{n}[(N_0/N)\xi_\nu].$$
Recall that we assume that the characteristic numbers of the cusp ∞ with respect to Γ_0 are zero. Then $(N_0/N)\xi_\nu$ is integral. We also mention that N_0/N is the index of $\Gamma_{0,\infty}$ in Γ_∞. Hence we get
$$\dim \mathcal{K}_x = \sum_{\nu=1}^{n}[\Gamma_\infty : \Gamma_{0,\infty}]\xi_\nu.$$
We recall that the characteristic numbers for a point $x \in X_\Gamma$ depend on (Γ, r, v). To point out this dependency we will write sometimes $\xi_\nu = \xi_\nu(r,v)$ and $\sigma(x) = \sigma_\Gamma(x,r,v)$ for their sum.

4.3 Lemma. Let $\pi : X_{\Gamma_0} \to X_\Gamma$ be the natural projection. We consider a point $x \in X_{\Gamma_0}$ which is the image of a cusp a. Let v be a multiplier system of weight r for Γ. We denote by $\sigma(a) = \sigma_\Gamma(a, r)$ the sum of the characteristic numbers. Then the formula

$$\dim \mathcal{K}_x = [\Gamma_a : \Gamma_{0,a}]\sigma(a), \qquad \mathcal{K} = \mathcal{M}_{\Gamma_0}(r, v)/\pi^* \mathcal{M}_\Gamma(r, v),$$

holds.

We use the notation

$$\pi : X_{\Gamma_0} \longrightarrow X_\Gamma$$

for the canonical map. We get a formula for the degree of $\mathcal{K}$.

4.4 Proposition. *The formula*

$$\deg \mathcal{K} = [\Gamma : \Gamma_0] \sum_{x \in X_\Gamma} \sigma(x)$$

holds. Here $\sigma(x) = \sigma_\Gamma(x, r, v)$ is the sum of the characteristic numbers at (a representative of) x.

We now get the link between the degrees of $\mathcal{M}_\Gamma(r, v)$ and $\mathcal{M}_{\Gamma_0}(r, v)$. The covering degree of $\pi : X_{\Gamma_0} \to X_\Gamma$ equals the index $[\Gamma : \Gamma_0]$. Using 4) from Theorem VI.4.3 we get the following formula.

$$[\Gamma : \Gamma_0] \deg \mathcal{M}_\Gamma(r, v) = \deg \mathcal{M}_{\Gamma_0}(r, v) - [\Gamma : \Gamma_0] \sum_{x \in X_\Gamma} \sigma(x).$$

The group Γ_0 can be chosen small enough such that the multiplier system is diagonal, that it acts fixed point free on $\mathbb{H}$, and that the characteristic numbers of the cusps are zero. Then $\mathcal{M}_{\Gamma_0}(r, v)$ is a direct sum of line bundles and we are reduced to the well-known case $V = \mathbb{C}$ which has been treated at various places in the literature. For sake of completeness we repeat shortly the argument. Since every line-bundle has a non-zero meromorphic section, there exists a non-zero meromorphic automorphic form f. We associate to f a divisor $D = (f)$ such that $\mathcal{O}_D$ is isomorphic to $\mathcal{M}_{\Gamma_0}(r, v)$. If $x \in X_{\Gamma_0}$ is the image of an inner point $a \in \mathbb{H}$, then $D(x)$ is the usual order of f at a. Let a be the cusp ∞. Since the characteristic numbers are zero, we can consider f as a holomorphic function in $q = e^{\frac{2\pi i}{N_0}\tau}$ and we define $D(x)$ to be the order of this function at $q = 0$. For an arbitrary cusp we use "transformation to ∞". It is easy to check that the order is independent of the choice of the transformation and even more that it depends only one the Γ_0-orbit of a. Let m be a natural number. Then one has $(f^m) = m(f)$ (since the characteristic numbers of the cusps vanish). We can take m such that mr is even and such that the multiplier system of f is trivial. Now we can compare with modular forms of weight two.

$$\deg \mathcal{M}_{\Gamma_0}(r, v) = \frac{r}{2} \deg \mathcal{M}_0(2).$$

§4. The computation of the degree, first method 143

We use that $\mathcal{M}_0^{\mathrm{cusp}}(2)$ is a canonical bundle and that the degree of the canonical bundle is $2g_0 - 2$. We obtain that the degree of $\mathcal{M}_0(2)$ is $2g_0 - 2 + h_0$ where h_0 denotes the number of cusp classes of Γ_0. Collecting together we obtain the following formula.

4.5 Proposition. *The formula*
$$\deg \mathcal{M}_\Gamma(r,v) = \frac{rn}{2[\Gamma:\Gamma_0]}(2g_0 - 2 + h_0) - \sum_{x \in X_\Gamma} \sigma_\Gamma(x,r,v).$$

holds. Here g_0 denotes the genus of X_{Γ_0}. The rank of v is denoted by n and h_0 denotes the number of cusp classes of Γ_0.

We want to express the formula above in data of the group Γ alone. For this we apply the Riemann–Hurwitz ramification formula VI.7.5 to the natural projection $\pi : X_{\Gamma_0} \to X_\Gamma$. It states
$$g_0 - 1 = \deg(\pi)(g-1) + \frac{1}{2}\sum_{a \in X_{\Gamma_0}}(\mathrm{Ord}(\pi,a) - 1).$$

The degree of $\pi : X_{\Gamma_0} \to X_\Gamma$ equals the index $[\Gamma : \Gamma_0]$. Since Γ_0 is a normal subgroup of Γ, the order at a point $a \in X_{\Gamma_0}$ depends only on its image $b \in X_\Gamma$. We denote this order by $e(b)$. So the ramification formula can be written as
$$\frac{g_0 - 1}{[\Gamma : \Gamma_0]} = (g-1) + \frac{1}{2}\sum_{b \in X_\Gamma}\left(1 - \frac{1}{e(b)}\right).$$

4.6 Lemma. *Let $\tilde{b} \in \mathbb{H}^*$ be a representative of $b \in X_\Gamma$. Then*
$$e(b) = [\Gamma_{\tilde{b}} : \Gamma_{0,\tilde{b}}].$$

In particular, if b is not a cusp, then $e(b) = \#\Gamma_{\tilde{b}}$ which is independent on the choice of Γ_0.

The number of inverse points of a given cusp class $b \in X_\Gamma$ is $[\Gamma_0 : \Gamma]/e(b)$. Hence we have
$$h_0 = \sum_{b \in X_\Gamma \text{ cusp}} \frac{[\Gamma_0 : \Gamma]}{e(b)}.$$

Now Proposition 4.7 can be rewritten in the following form.

4.7 Proposition. *The degree formula can be written as*
$$\deg \mathcal{M}_\Gamma(r,v) = rn\left(g - 1 + \frac{h}{2} + \frac{1}{2}\sum_{b \in X_\Gamma \text{ not cusp}}\left(1 - \frac{1}{e(b)}\right)\right) - \sum_{x \in X_\Gamma} \sigma_\Gamma(x,r,v).$$

5. The dimension formula

In this section we assume that the assumption 4.1 is true.

The Riemann–Roch formula states

$$\chi(\mathcal{M}_\Gamma(r,v)) = \deg(\mathcal{M}_\Gamma(r,v)) + \mathrm{Rank}(\mathcal{M}_\Gamma(r,v))(1-g).$$

We are more interested in the spaces of automorphic forms

$$[\Gamma, r, v] := H^0(X_\Gamma, \mathcal{M}_\Gamma(r,v)).$$

The Serre dual space is the subspace of cusp forms of $[\Gamma, v', 2-r]$. In the case $r > 2$ this space vanishes.

In the case $r = 2$ there is a difference depending on the fact whether Γ has a cusp or not. We first mention that v is a representation (homomorphism) in this case. Modular forms of weight 0 are constants. Hence $[\Gamma, v', 0]$ is just the space of v'-invariants of V. This is isomorphic to the space of v-invariants of V. We denote the space of invariants by V^v. Since constant cusp forms are zero if there is a cusp, we obtain the following dimension formula.

5.1 Theorem. *In the case $r > 2$ we have*

$$\dim[\Gamma, r, v] = rn\left(g - 1 + \frac{h}{2} + \frac{1}{2}\sum_{b \in X_\Gamma \text{ not cusp}} \left(1 - \frac{1}{e(b)}\right)\right)$$
$$+ n(1-g) - \sum_{x \in X_\Gamma} \sigma(x).$$

Here g is the genus of X_Γ. The dimension of V is n and $\sigma(x) = \sigma_\Gamma(x, r, v)$ is the sum of the characteristic numbers. The number $e(b)$ is the order of the stabilizer of a representative of b in $\mathbb{H}$.

Supplement. *When Γ has a cusp, then this formula remains true in the case $r = 2$. Otherwise one has to add $\dim V^v$ to the right hand side.*

We denote by $[\Gamma, r, v]_0$ the subspace of cusp forms of $[\Gamma, r, v]$. This is the space of global sections of the sheaf $\mathcal{M}^{\mathrm{cusp}} = \mathcal{M}^{\mathrm{cusp}}(\Gamma, r, v)$. Since the quotient $\mathcal{M}/\mathcal{M}^{\mathrm{cusp}}$ is a skyscraper sheaf, we have

$$\chi(\mathcal{M}) - \chi(\mathcal{M}^{\mathrm{cusp}}) = \sum_{x \in X_\Gamma, \text{ cusp}} \dim(\mathcal{M}_x/\mathcal{M}_x^{\mathrm{cusp}}).$$

Recall that $\mathcal{M}_x$ is given by Fourier series with summation over integers m such that $m + \xi_\nu(r) \geq 0$ and, in the subspace $\mathcal{M}_x^{\mathrm{cusp}}$, the summation is restricted to $m + \xi_\nu(r) > 0$. There is only a difference if the characteristic number ξ_ν is zero. We see

$$\chi(\mathcal{M}^{\mathrm{cusp}}) = \chi(\mathcal{M}) - \sum_{x \in X_\Gamma, \text{ cusp}} \#\{\nu;\ \xi_\nu(r) = 0\}.$$

5.2 Remark. *Assume that Γ has cusp ∞. The number*

$$\#\{\nu;\ \xi_\nu = 0\}$$

equals the dimension of the subspace of invariants of V under the transformations $J(\gamma, \tau)$, $\gamma \in \Gamma_\infty$. (These transformations do not depend on τ.)

Finally we formulate the dimension formula for the space of cusp forms. In the case of weight 2 we have to be careful, since

$$\chi(\mathcal{M}^{\mathrm{cusp}}(\Gamma, v, 2)) = \dim[\Gamma, v, 2]_0 - \dim[\Gamma, v', 0].$$

In the case of an even weight, v is a representation and $[\Gamma, v', 0]$ can be identified with the space of invariants of v'. We obtain the following result.

5.3 Theorem. *Assume that Γ has at least one cusp. In the case $r \geq 2$ the dimension of the space of cusp forms is*

$$\dim[\Gamma, r, v]_0 = \dim[\Gamma, r, v] - \sum_{x \in X_\Gamma,\ \mathrm{cusp}} \#\{\nu;\ \xi_\nu(r) = 0\}.$$

6. The full modular group

We specialize the dimension formula to the group $\Gamma = \mathrm{SL}(2, \mathbb{Z})/\pm$. As usual it acts on the upper half plane by $(a\tau + b)(c\tau + d)^{-1}$. In this case $g = 0$ and $h = 1$. We have two classes of elliptic fixed points of order $e = 2$ resp. $e = 3$. In the dimension formula we get

$$\sum_{b \in X_\Gamma \text{ not cusp}} \left(1 - \frac{1}{e(b)}\right) = \left(1 - \frac{1}{2}\right) + \left(1 - \frac{1}{3}\right) = \frac{7}{6}.$$

So the dimension formula gives

$$\dim[\Gamma, r, v] = \frac{rn}{12} + n - \sum_{x \in X_\Gamma} \sigma(x).$$

We use the usual generators

$$T = \begin{pmatrix} 1 & 1 \\ 0 & 1 \end{pmatrix},\ \begin{pmatrix} 0 & -1 \\ 1 & 0 \end{pmatrix}.$$

Representatives of the elliptic fixed points are i and $\zeta_3 = -1/2 + i\sqrt{3}/2$. The elements in their stabilizers which correspond to the rotation with factor $e^{2\pi i/e}$ can be computed easily as S resp. $(ST)^{-1}$.

Let A be a complex matrix of finite order. The eigenvalues are roots of unity which we can write in the form
$$\lambda = \exp(2\pi i\alpha) \text{ with } 0 \leq \alpha < 1.$$
We use the notation
$$\alpha(A) = \sum_\lambda \alpha,$$
where λ runs through all eigenvalues (counted with multiplicity).

The contributions of the characteristic numbers in the dimension formula can be written as
$$\sum_{x \in X_\Gamma} (\xi_1(r) + \cdots + \xi_n(r)) = \alpha(J(S, i)) + \alpha(J((ST)^{-1}, \zeta_3)) + \alpha(J(T, \cdot)).$$
(The function $J(T, \tau)$ is independent of τ.)

6.1 Theorem. *In case of the full modular group the dimension formula is valid for $r \geq 2$ (including $r = 2$) and reads as*
$$\dim[\Gamma, \varrho, r] = \frac{rn}{12} + n - \alpha(J(S, i)) - \alpha(J((ST)^{-1}, \zeta_3)) - \alpha(J(T, \cdot)).$$
For the subspace of cusp forms one has
$$\dim[\Gamma, r, v]_0 = \dim[\Gamma, r, v] - \dim V^{J(T, \cdot)} + \begin{cases} 0 & \text{if } r > 2, \\ \dim V^\varrho & \text{if } r = 2. \end{cases}$$

We treat a simple example just to get a feeling how the formula works. The weight r is assumed to be even and we consider the case of a trivial multiplier system. This means $J(\gamma, \tau) = (c\tau + d)^{r/2}$. So we get
$$J(S, i) = e^{2\pi i r/4}$$
and
$$J((ST)^{-1}, \zeta_3) = e^{2\pi i r/6}.$$
The sum of both is
$$\begin{cases} 0 & \text{for } r \equiv 0 \mod 12, \\ 7/6 & \text{for } r \equiv 2 \mod 12, \\ 1/3 & \text{for } r \equiv 4 \mod 12, \\ 1/2 & \text{for } r \equiv 6 \mod 12, \\ 2/3 & \text{for } r \equiv 8 \mod 12, \\ 5/6 & \text{for } r \equiv 10 \mod 12. \end{cases}$$
Using the table above, one gets
$$\begin{cases} \left[\frac{r}{12}\right] & \text{if } r \equiv 2 \mod 12, \\ \left[\frac{r}{12}\right] + 1 & \text{else.} \end{cases}$$
This formula is true for all even $r > 0$ (also for $r = 2$).

7. The metaplectic group

There is a different way to express multiplier systems using the metaplectic group. We recall this concept briefly in the case of half-integral weight. The metaplectic group
$$\mathrm{Mp}(2,\mathbb{R}) \longrightarrow \mathrm{SL}(2,\mathbb{R})$$
can be described as the set of all pairs (M, J), where $M = \begin{pmatrix} a & b \\ c & d \end{pmatrix} \in \mathrm{SL}(2,\mathbb{R})$, and where $J = \sqrt{c\tau + d}$ is one of the two holomorphic square roots of the function $c\tau + d$ on the upper half plane $\mathbb{H}$. The group law is
$$(M, \sqrt{c\tau + d})(M', \sqrt{c'\tau + d'}) = (MM', \sqrt{c'\tau + d'}\sqrt{cM'\tau + d}).$$
One knows that $\mathrm{Mp}(2, \mathbb{Z})$ is generated by
$$T = \left(\begin{pmatrix} 1 & 1 \\ 0 & 1 \end{pmatrix}, 1\right), \quad S = \left(\begin{pmatrix} 0 & -1 \\ 1 & 0 \end{pmatrix}, \sqrt{\tau}\right), \quad \mathrm{Re}\,\tau > 0,$$
and that the relations
$$S^2 = (ST)^3 = Z, \quad Z = \left(\begin{pmatrix} -1 & 0 \\ 0 & -1 \end{pmatrix}, \mathrm{i}\right), \quad Z^4 = 1$$
are defining ones.

Let
$$\varrho : \mathrm{Mp}(2, \mathbb{Z}) \longrightarrow \mathrm{GL}(V)$$
be a representation of $\mathrm{Mp}(2, \mathbb{Z})$ on some finite dimensional complex vector space. Let r be an integer or a half integer ($2r \in \mathbb{Z}$). An (entire) modular form of weight r with respect to ϱ is holomorphic function $f : \mathbb{H} \to V$ with the transformation law
$$f(M\tau) = \sqrt{c\tau + d}^{\,2r} \varrho(M) f(\tau) \quad \text{for all} \quad (M, \sqrt{c\tau + d}) \in \mathrm{Mp}(2, \mathbb{Z})$$
and such that f is bounded for $\mathrm{Im}\,\tau \geq 1$.

We denote by $[\mathrm{Mp}(2, \mathbb{Z}), r, \varrho]$ the space of all entire modular forms.

Let $V_0 \subset V$ be the subspace on which $\varrho(-E, \mathrm{i})$ (E denotes the unit matrix) acts by multiplication with $e^{-\pi\mathrm{i}r} = \mathrm{i}^{-2r}$. It is quite clear that the values of f are contained in V_0 and that V_0 is invariant under $\mathrm{Mp}(2, \mathbb{Z})$. Let $\gamma \in \mathrm{SL}(2, \mathbb{Z})/\pm$ a modular transformation. We choose a pre-image $(M, \sqrt{c\tau + d}) \in \mathrm{Mp}(2, \mathbb{C})$ and define the operator
$$J(\gamma, \tau)a = \sqrt{c\tau + d}^{\,2r} \varrho(M)a \quad \text{for} \quad a \in V_0.$$
This is independent of the choice of the pre-image (since we restrict to V_0). For trivial reason
$$J(\gamma, \tau)a'(\tau)^{r/2}$$
is a V_0-valued multiplier system v and Assumption 4.1 is satisfied. The space of automorphic forms $[\mathrm{SL}(2, \mathbb{Z})/\pm, r, v]$ of weight r with respect to this multiplier system coincides with $[\mathrm{Mp}(2, \mathbb{Z}), \varrho, r]$. Hence the dimension formula gives the following result.

7.1 Theorem. Let $\varrho : \mathrm{Mp}(2,\mathbb{Z}) \to \mathrm{GL}(V)$ be a representation on a finite dimensional vector whose image is finite. Let V_0 be the biggest subspace of V where $\varrho(-E,\mathrm{i})$ acts by multiplication with $e^{-\pi\mathrm{i}r}$. We denote by d the dimension of V_0. Then one has for $r \geq 2$ (including $r = 2$)

$$\dim[\mathrm{Mp}(2,\mathbb{Z}),\varrho,r] = \frac{rd}{12} + d - \alpha(e^{\pi\mathrm{i}r/2}\varrho(S)) - \alpha\!\left(\left(e^{\pi\mathrm{i}r/3}\varrho(ST)\right)^{-1}\right) - \alpha(\varrho(T)).$$

The invariants α have to be taken with respect to the action on V_0.

For the subspace of cusp forms one has

$$\dim[\mathrm{Mp}(2,\mathbb{Z}),\varrho,r]_0 = \dim[\mathrm{Mp}(2,\mathbb{Z}),\varrho,r] - \dim V_0^{\varrho(T)} + \begin{cases} 0 & \text{if } r > 2, \\ \dim V_0^{\varrho} & \text{if } r = 2. \end{cases}$$

The operator $e^{\pi\mathrm{i}r/2}\varrho(S)$ (considered on V_0) has order 2 and $\left(e^{\pi\mathrm{i}r/3}\varrho(ST)\right)^{-1}$ has order 3. Their α-invariants can be computed very easily be means of the following lemma.

7.2 Lemma. Let A be a $d \times d$-matrix. We have

$$\alpha(A) = \begin{cases} \dfrac{d}{4} - \dfrac{\mathrm{tr}(A)}{4} & \text{if } A^2 = E, \\[2mm] \dfrac{d}{3} - \dfrac{1}{3}\mathrm{Re}(\mathrm{tr}(A^{-1})) + \dfrac{1}{3\sqrt{3}}\mathrm{Im}(\mathrm{tr}(A^{-1})) & \text{if } A^3 = E. \end{cases}$$

… §8. Chern forms …

Appendix. Generalizations of the dimension formula

We sketch another method for the computation of the degree which gives the dimension formula for arbitrary *real* weight and arbitrary multiplier systems in the sense of Definition 2.1. This method is more involved and uses hermitian metrics on bundles and their Chern forms. It is related to the Gauss–Bonnet formula.

8. Chern forms

We describe a different method for the computation of the degree of a line bundle. This uses more general theory which we do not want to develop here, since the cases we could treat so far are sufficient usually. So we will keep very short and expect a reader who has some experience and is willing to fill the gaps himself.

We want to consider *hermitian forms* on a vector bundle $\mathcal{M}$,

$$\mathcal{M} \times \mathcal{M} \longrightarrow \mathcal{C}_X^\infty.$$

By definition, this is a family of maps

$$\langle \cdot, \cdot \rangle : \mathcal{M}(U) \times \mathcal{M}(U) \longrightarrow \mathcal{C}_X^\infty(U) \qquad (U \subset X \text{ open}),$$

compatible with restriction such that the rules

$$\langle s_1 + s_2, t \rangle = \langle s_1, t \rangle + \langle s_2, t \rangle, \quad \langle fs, t \rangle = f\langle s, t \rangle, \quad \langle s, t \rangle = \overline{\langle t, s \rangle}$$

are satisfied. Here s, s_1, s_2, t are sections of $\mathcal{M}$ and f is a holomorphic function. The expressions $\langle s, s \rangle$ are real functions. We call the hermitian form *positive semidefinite* if $\langle s, s \rangle$ is nowhere negative.

We want to define what it means that a hermitian form is *positive definite*. For this we have to introduce the value $s(a)$ of a section $s \in \mathcal{M}(U)$ at a point $a \in U$. We consider the $\mathbb{C}$-vector space

$$\mathcal{M}(a) := \mathcal{M}_a \otimes_{\mathcal{O}_{X,a}} \mathbb{C}.$$

Here $\mathbb{C}$ is considered as module over the ring $\mathcal{O}_{X,a}$ through the natural homomorphism $\mathcal{O}_{X,a} \to \mathbb{C}$ (evaluation at a). If $\mathcal{M}$ has rank n, then $\mathcal{M}(a)$ is an n-dimensional vector space. We define $s(a)$ to be the image of s_a under the natural map

$$\mathcal{M}_a \longrightarrow \mathcal{M}(a), \quad s \longmapsto a \otimes 1.$$

In the special case $\mathcal{M} = \mathcal{O}_X$ we can identify $\mathcal{O}_X(a)$ and $\mathbb{C}$ and then $f(a)$ for a section $f \in \mathcal{O}_X(U)$ is just the value of the function f at the point a.

Now we can define when a positive semidefinite hermitian form is called definite. The condition is that for each section $s \in \mathcal{M}(U)$ and for each point $a \in U$ we have

$$\langle s, s \rangle(a) = 0 \Longrightarrow s(a) = 0.$$

We are mainly interested in hermitian forms on line bundles. In this case the definition of positive definiteness is still simpler. So let $\mathcal{L}, \langle \cdot, \cdot \rangle$ be a line bundle which has been equipped with a positive definite hermitian form. Let $U \subset X$ an open subset such that $\mathcal{L}|U$ is trivial and let $s \in \mathcal{L}(U)$ be a generator. This means $\mathcal{L}(V) = s\mathcal{O}(V)$ for all open $V \subset U$. We consider the function $\langle s, s \rangle$. Positive definiteness means that this function is positive everywhere. Hence we can consider the differential form of degree two $\partial\bar{\partial} \log \langle s, s \rangle$. We claim that this differential form is independent on the choice of the generator s and hence all these differential forms for varying U glue to a global differential form on X.

8.1 Proposition. *Let $\mathcal{L}, \langle \cdot, \cdot \rangle$ be a line bundle which has been equipped with a positive definite hermitian form. There exists a unique differential form $\omega \in A^2(X)$ such that for each generator s of $\mathcal{L}|U$ on any open subset the equality*

$$\omega|U = \frac{1}{2\pi\mathrm{i}} \partial\bar{\partial} \log \langle s, s \rangle$$

holds.

Proof. If s is a generator, then each other generator is of the form fs with a holomorphic function without zeros. The operator $\bar{\partial}$ kills holomorphic functions. Similarly the operator ∂ kills the complex conjugate of a holomorphic function. Another rule which can be verified is $\partial\bar{\partial} h = -\bar{\partial}\partial h$ for differentiable functions h. Hence $\partial\bar{\partial}$ kills holomorphic functions and complex conjugates of holomorphic functions. Locally, the function $\log(f\bar{f})$ is the sum of a holomorphic function and the complex conjugate of a holomorphic function. This shows

$$\partial\bar{\partial} \log(f\bar{f}) = 0.$$

We obtain

$$\bar{\partial}\partial \log(f\bar{f}h) = \bar{\partial}\partial \log h$$

for positive differentiable functions h. So the differential form is independent on the choice of the generator s. □

We call ω a *Chern form* of the line bundle $\mathcal{L}$. It depends on the choice of the hermitian form. Chern forms can be used to compute the degree of a line bundle.

§8. Chern forms

8.2 Theorem. *Let ω be a Chern form of a line bundle X. Then the formula*

$$\deg(\mathcal{L}) = \int_X \omega$$

holds.

We indicate the proof. First one shows that the integral is independent of the choice of the Chern form. Then one shows that the theorem is true for a covering $X \to Y$ if it is true for Y. So it is sufficient to verify it for the Riemann sphere and a point divisor, for example the divisor (∞). In this case everything can be made explicit. □

As a warm-up we apply Theorem 8.2 in the simplest case that $X = X_\Gamma$, where Γ has no elliptic and parabolic fixed points. Then $\mathbb{H}/\Gamma$ is compact and the natural projection $\mathbb{H} \to \mathbb{H}/\Gamma$ is locally biholomorphic. Holomorphic (resp. differentiable) functions on an open subset of $\mathbb{H}/\Gamma$ are in one-to-one correspondence to Γ-invariant holomorphic (resp. differentiable) functions on the inverse image in $\mathbb{H}$. We consider now for even r the trivial multiplier system v. Hence we study the bundle $\mathcal{M}(r)$ which describes the transformation law

$$f(\gamma\tau) = (c\tau + d)^{2r} f(\tau), \quad \gamma \leftrightarrow \pm \begin{pmatrix} a & b \\ c & d \end{pmatrix}.$$

If f, g are two functions with this transformation formula, then

$$\langle f, g \rangle := f(\tau)\overline{g(\tau)} \operatorname{Im} \tau^r$$

is Γ invariant and defines a differentiable function on some open subset of $\mathbb{H}/\Gamma$. This gives a positive definite hermitian form on $\mathcal{M}(r)$. We compute its Chern form. Since the operators ∂ and $\bar{\partial}$ are compatible with pull-back with respect to holomorphic mappings, we obtain that the Chern form can be identified with the Γ-invariant differential form

$$\partial\bar{\partial} \log(y^r) = \frac{ri}{2} \frac{dx \wedge dy}{y^2}.$$

Now we have a new formula for the degree of $\mathcal{M}(r)$. In Proposition we obtained the expression $2r(g-1)$. Comparing it with the new computation we obtain (from Theorem 8.2)

$$\frac{1}{2\pi} \int_{\mathbb{H}/\Gamma} \frac{dxdy}{y^2} = 2g - 2.$$

This is a special case of the *Gauss–Bonnet formula* from differential geometry.

9. The computation of the degree, second method

Now we allow that Γ contains elliptic and parabolic fixed points. We make some comments about integration on $\mathbb{H}/\Gamma$. Let $D \subset \mathbb{C}$ be an open domain and let Γ be a group of biholomorphic transformations, acting discontinuously. Let $\pi : D \to D/\Gamma$ be the natural projection. There exists a discrete subset $S \subset D$ such that Γ acts without fixed points on $D - S$. Let ω be a nowhere negative top differential form on $X = (D - S)/\Gamma$. This defines a measure on X. This measure extends to a measure on D/Γ such that the image of S gets measure 0. In particular, we can talk about the volume of D/Γ which may be finite or not. The volume can be computed with the help of a fundamental domain. We briefly explain what this means. We consider subsets $F \subset D$ with the following property. We assume that F is the closure of its interior and that the boundary has measure 0. We say that F is a *fundamental set* if D is the union of all $\gamma(D)$, $\gamma \in \Gamma$. It is called a *fundamental domain* if, in addition, two different points in the interior of F are not equivalent mod Γ. For a fundamental domain F the formula

$$\int_{\mathbb{H}/\Gamma} \omega = \int_F \pi^*\omega$$

holds for trivial reason. If F is only a fundamental set, then still the inequality "$\le$" holds.

9.1 Lemma. *Let Γ be a subgroup of $\mathrm{Aut}(\mathbb{H})$ such that its inverse image in $\mathrm{SL}(2,\mathbb{R})$ is discrete and such that $\mathbb{H}^*/\Gamma$ is compact. The volume of $\mathbb{H}/\Gamma$ with respect to the measure $dxdy/y^2$ is finite.*

Proof. We construct a suitable fundamental set. Let ∞ be a cusp of Γ. By a *cusp sector* at ∞ we understand a domain of the form $\mathrm{Im}\,\tau \ge C, |\mathrm{Re}\,\tau| \le C'$ where C, C' are positive. The notion of a cusp sector at an arbitrary cusp is defined in the usual way by transforming it to ∞. From the compactness of $\mathbb{H}^*/\Gamma$ one can deduce that there exists a fundamental set which is the union of a compact set in $\mathbb{H}$ and finitely many cusp sectors ([Fr2], Proposition I.1.15). So we are reduced to the statement that the cusp sector at ∞ has finite volume with respect to $dxdy/y^2$. This is easy to check. □

We want now to construct a hermitian form on $\mathcal{M}(r,v)$ in the general (scalar valued) case. Of course we could try to take

$$\langle f,g\rangle := f(\tau)\overline{g(\tau)}(\mathrm{Im}\,\tau)^r$$

also in this case. But this causes problems at the elliptic fixed points and at the cusps. At the elliptic fixed points the following problem arises. The transformation law $f(\gamma\tau) = J(\gamma,\tau)f(\tau)$ implies that $f(a) = 0$ if a is the fixed point of a γ with the property $J(\gamma,a) \ne 1$. Hence $\langle f,f\rangle$ would vanish at such

§9. The computation of the degree, second method

fixed points. But this would contradict to the definiteness. The situation at the cusps is still worse. Let ∞ be cusp, then $\langle f, f \rangle (\operatorname{Im} \tau)^r$ can not be expected to extend to a C^∞-function at $[\infty]$. Hence we have to modify the definition of $\langle f, g \rangle$ close to an elliptic fixed point or to a cusp.

We modify the definition of the hermitian metric as follows. We will construct a certain Γ-invariant positive differentiable function ϕ on $\mathbb{H} - S$ where S denotes the set of elliptic fixed points. With this function we will modify the definition of the hermitian metric as follows. Let $U \subset X_\Gamma$ be open and let $\tilde{U}$ its inverse image in $\mathbb{H}$. Then we will define

$$\langle f, g \rangle := f(\tau)\overline{g(\tau)}(\operatorname{Im} \tau)^r \cdot \phi(\tau).$$

This is a Γ-invariant function on $\tilde{U}$ and can be considered as a function on $\tilde{U}/\Gamma$. This is an open subset of U but cusp classes may be missing. So we will have to make the choice of ϕ in such a way that $\langle f, g \rangle$ extends to a continuous function on the whole U and even more, this function should be differentiable. And we want to have the that the modified $\langle \cdot, \cdot \rangle$ is positive definite.

Assume that such a function has been constructed. Then we get the formula

$$\deg \mathcal{M}(r, v) = \frac{1}{2\pi} \int_{\mathbb{H}/\Gamma} \frac{dx dy}{y^2} + \frac{1}{2\pi i} \int_{\mathbb{H}/\Gamma} \partial \bar{\partial} \log \phi(\tau).$$

To construct a suitable ϕ, we choose for each point $x \in X_\Gamma$, which is the image of an elliptic fixed point or a cusp, a small open neighborhood $U(x)$ such that these are pairwise disjoint. We will construct ϕ such that the support of $\phi(\tau) - 1$ is contained in the union of the $U(x)$. The construction of ϕ can be done in each $U(x)$ separately.

We treat first the case such that $x = [a]$ is the image of an elliptic fixed point a. We use the chart $w = (\tau - a)(\tau - \bar{a})^{-1}$ as has been described in Remark and Definition 2.3. The neighborhood is defined by $|w| < r$ where r is sufficiently small. We have to make use of the characteristic number ξ, $0 \leq \xi < 1$. From its definition follows that the function

$$s = w^{e\xi} \quad (e = \#\Gamma_a)$$

defines a section of $\mathcal{M}(r, v)$ in $U(a)$. It is clear that this is a generating section. The expression $s\bar{s}(\operatorname{Im} \tau)^r$ vanishes at a. This is the reason why we have to introduce the modifying function φ. Close to a it should agree with $(\bar{w}w)^{-e\xi}$. So we have to demand:

For an elliptic fixed point a the function ϕ should be defined such that it is an everywhere positive differentiable function, such that $\phi(\tau) - 1$ has compact support on $U(a)$, and such that it agrees with $(\bar{w}w)^{-e\xi}$ on a small neighborhood of a.

The elliptic fixed point gives an extra contribution to the degree, namely

$$\int_{U(a)/\Gamma_a} \partial\bar\partial \log \phi.$$

This integral agrees with

$$\frac{1}{e}\int_{U(a)} \partial\bar\partial \log \phi.$$

It can be computed as follows.

9.2 Lemma. *Let $r > 0$ and let ϕ be an everywhere positive differentiable function on the disk $|w| < r$, such that $\phi - 1$ has compact support. Assume that ϕ agrees on a small neighborhood of the origin with the function $(\bar w w)^\xi$ for some real number ξ. Then the formula*

$$\frac{1}{2\pi\mathrm{i}}\int_{|w|<r} \partial\bar\partial \log \phi = \xi$$

holds.

Proof. The prove is an application of the Stokes formula. First we notice that $\partial\bar\partial = d\bar\partial$. This follows from the rule $\bar\partial^2 = 0$. Then we consider for small ε the domain $\varepsilon < |w| < r$. ("Small" means that $2\varepsilon < r$ and that ϕ agrees with $\bar w w^\xi$ for $|w| < 2\varepsilon$.) To this domain we can apply the formula of Stokes.

$$\int_{\varepsilon<|w|<r} \partial\bar\partial \log \phi = \oint_{|w|=\varepsilon} \bar\partial \log \phi.$$

The circle line $|w| = \varepsilon$ has to be oriented such that the integration area is on its left. This means that we have to take the clockwise orientation (opposite to the usual mathematical orientation). We set $w = u + \mathrm{i}v$ and compute

$$\frac{\partial}{\partial \bar w}\log \phi = \frac{1}{2}\Big(\frac{\partial}{\partial u} + \mathrm{i}\frac{\partial}{\partial v}\Big)\log(u^2+v^2)^\xi = \frac{\xi}{w}.$$

Hence we obtain

$$\oint_{|w|=\varepsilon} \bar\partial \log \phi = \xi \oint_{|w|=\varepsilon} \frac{d\bar w}{\bar w}.$$

To compute the integral of $d\bar w/\bar w$, we use the general formula for line integrals $\overline{\int \eta} = \int \bar\eta$ and we use that the integral of dw/w is $-2\pi\mathrm{i}$ due to the "wrong" orientation of the circle line. The integral is independent of ε. So there is no problem to take the limit $\varepsilon \to 0$ which completes the proof of Lemma 9.2. ☐

One should compare this with the degree formula 4.7. There is a term $\sigma(\xi)$ whose occurrence – for elliptic fixed points – now finds a new explanation.

§9. The computation of the degree, second method

Next we treat the case that a is cusp. We can assume that it is the cusp ∞. Then we take the neighborhood in the form $\operatorname{Im}\tau > C$ with sufficiently big C. The stabilizer Γ_∞ is generated by a substitution $\tau \mapsto \tau + N$. The function $q^{\xi/N}$ (recall $q^a := e^{2\pi i a\tau}$) is a generating section. Hence in this case it looks natural to define the compensating function $\phi(\tau)$ for $\operatorname{Im}\tau > C$ such that it is everywhere positive, periodic, and such that for very big $\operatorname{Im}\tau$ it agrees with $\bar{q}q^{-\xi/N}$ and, finally, that $\phi(\tau) - 1$, considered as a function of $q^{1/N}$, has compact support. But this is not enough. The reason is that there occurs also the function y^r which is increasing for $y \to \infty$. This function has to by compensated too. Hence the correct choice of ϕ is as follows.

For the cusp ∞, the function ϕ should be defined such that it is a differentiable, everywhere positive, periodic function for $\operatorname{Im}\tau > C$, such that $\phi(\tau) - 1$, considered as function of $q^{1/N}$, has compact support, and such that it agrees with

$$(\bar{q}q)^{-\xi/N} y^{-r}$$

for very big $\operatorname{Im}\tau$.

The computation is now essentially the same as in the case of an elliptic fixed point. There only occurs an extra contribution

$$\oint_{|q|=\varepsilon} \bar{\partial} y^r.$$

It is easy to evaluate this integral and to show that it tends to 0 for $\varepsilon \to 0$. Collecting together, we obtain (in the scalar valued case) the degree formula

$$\deg \mathcal{M}(r,v) = \frac{r}{4\pi} \int_{\mathbb{H}/\Gamma} \frac{dxdy}{y^2} - \sum_a \xi(a).$$

For example, for even r and trivial v, it must coincide with the degree formula 4.7 that we derived with our first method. Comparing them gives the following variant of the Gauss–Bonnet formula.

9.3 Theorem.

$$\frac{1}{2\pi} \int_{\mathbb{H}/\Gamma} \frac{dxdy}{y^2} = 2g - 2 + h + \sum_{a \in \mathbb{H}/\Gamma \text{ (no cusp)}} \left(1 - \frac{1}{e(a)}\right).$$

As an example, we can take the modular group $\mathrm{SL}(2,\mathbb{Z})/\pm$. The volume of the fundamental domain $|\operatorname{Re}\tau| \leq 1$, $|\tau| \geq 1$ computes as $\pi/3$. There is one cusp class, and there are two classes of elliptic fixed points of order 2 and 3. The genus is zero.

156   Chapter VIII. Dimension formulae for automorphic forms

The vector valued case

We want to extend the degree formula to the case of a vector valued multiplier system (r, v). For this we want to use the fact that a vector bundle $\mathcal{M}$ and its determinant (highest exterior power) $\bigwedge^n \mathcal{M}$ have the same degree (Remark VI.4.6). The determinant of $\mathcal{M}(r, v)$ is related to the scalar valued multiplier system $(\det v, rn)$. To explain this, we assume that $V = \mathbb{C}^n$. The vector valued modular forms are written as columns. Then we can consider the $\mathcal{O}_{X_\Gamma}$ multilinear map

$$\mathcal{M}(r,v)^n \longrightarrow \mathcal{M}(\det v, rn), \quad (f_1, \ldots, f_n) \longmapsto \det(f_1, \ldots, f_n).$$

This is an alternating multilinear map of $\mathcal{O}_{X_\Gamma}$-modules which induces a map $\bigwedge^n \mathcal{M}(r, v) \to \mathcal{M}(\det r, vn)$. This map is injective. Hence we obtain an exact sequence

$$0 \longrightarrow \bigwedge^n \mathcal{M}(r,v) \longrightarrow \mathcal{M}(\det r, vn) \longrightarrow \mathcal{K} \longrightarrow 0.$$

We compute $\mathcal{K}_a$. The computation will show that this is 0 if a is not the image of an elliptic fixed point or a cusp. We explain the computation in the case that a is the image of the cusp ∞ and leave the other cases to the reader. We can assume that $v(t_N)$ is diagonalized for the generating translation. Then $\mathcal{M}_a(r, v)$ is generated by $q^{\xi_\nu/N} e_\nu$. Taking the determinant, we get the generator

$$q^{\sigma(a)/N}, \quad \sigma(a) = \xi_1 + \ldots + \xi_n.$$

But the generator for the line bundle $\mathcal{M}(\det r, vn)$ is

$$q^{(\sigma(a) - [\xi_1 + \ldots + \xi_n])/N}.$$

Here we use the Gauss bracket $[x] = \max\{\nu \in \mathbb{Z}; \ \nu \leq x\}$. Notice that $x - [x]$ is the representative of the coset $x + \mathbb{Z}$ in $[0, 1)$. So we see

$$\dim \mathcal{K}_a = [\xi_1 + \cdots + \xi_n].$$

As we mentioned, a similar computation works for all points a. This shows

$$\deg(\mathcal{K}) = \sum_a [\sigma(a)].$$

The short exact cohomology sequence gives

$$\deg \mathcal{M}(r, v) = \deg \mathcal{M}(\det r, vn) - \sum_a [\sigma(a)].$$

Now we can insert the degree formula for the scalar valued case.

§9. The computation of the degree, second method

9.4 Theorem. *The degree formula that we derived under the special assumptions for (r, v) in Proposition 4.7 holds for arbitrary real weight and multiplier system v. In particular, the dimension formula in Theorem 5.1*

$$\dim[\Gamma, r, v] = rn\left(g - 1 + \frac{h}{2} + \frac{1}{2}\sum_{b \in X_\Gamma \text{ not cusp}}\left(1 - \frac{1}{e(b)}\right)\right)$$
$$+ n(1 - g) - \sum_{x \in X_\Gamma} \sigma(x)$$

is true for $r > 2$ (and with the same correction in the case $r = 2$) in the general case.

References

[Bo1] Borcherds, R.: *The Gross–Kohnen–Zagier theorem in higher dimensions,* Duke Math. J. **97**, no. 2, 219–233 (1999), Correction: Duke Math. J. **105**, no. 1, 183–184 (2000)

[Bo2] Borcherds, R.: *Reflection groups of Lorentzian lattices,* Duke Math. J. **104**, no. 2, 319–366 (2000)

[FB] Freitag, E., Busam, R.: *Complex Analysis,* Springer–Universitext, Springer-Verlag Berlin Heidelberg New-York (2005)

german edition: *Funktionentheorie 1,* Springer–Lehrbuch, 4. Auflage, Springer-Verlag Berlin Heidelberg New-York (2006)

[Fi] Fischer, J.: *An approach to the Selberg trace formula via the Selberg zeta-function,* Lecture Notes in Mathematics, **1253**, Springer Verlag Berlin Heidelberg New York (1987)

[Fo] Forster, O.: *Lectures on Riemann Surfaces,* Graduate Texts in Mathematics **81** Springer-Verlag Berlin Heidelberg New-York (1999),

german edition: *Riemannsche Flächen* Heidelberger Taschenbücher **184** Springer-Verlag Berlin Heidelberg New-York (1977)

[Fr1] Freitag, E.: *Complex Analyis 2,* Springer Universitext, Springer-Verlag Berlin Heidelberg New-York (2011)

german edition: *Funktionentheorie 2,* Springer-Lehrbuch, Springer-Verlag Berlin Heidelberg New-York (2009)

[Fr2] Freitag, E.: *Hilbert modular forms,* Springer-Verlag Berlin Heidelberg New-York (1990)

[GR] Gunning, R.C., Rossi, H.: *Analytic functions of several complex variables,* Prentice-Hall, Englewood Cliffs, NJ. (1965)

[Iw] Iwaniec, H.: *Spectral Methods of Automorphic Forms,* American Mathematical Society, Providence, Rhode Island, 2. ed. (2002)

[Qu] von Querenburg, B.: *Mengentheoretische Topologie,* Springer-Verlag Berlin Heidelberg New-York (1979)

[Sk] Skoruppa, N,P.: *Jacobi forms and modular forms of half-integral weight,* Ph.D. thesis (Dr. rer. nat.) University of Bonn (1985)

Index

Abel's theorem 108
abelian function 117f
— group 19
Abel–Jacobi 105
— map 105
acyclic 38
algebraic function 62
analytic atlas 52
— chart 52
— continuation 61
— manifold 118
arc components 13
arcwise connected 13
associated flabby sheaf 31
automorphic form 121, 127
automorphy factor 128

Ball 15
Banach space 16f
bimeromorphic 116
Borcherds 121
boundary point 9

Canonical divisor 93f
— flabby resolution 34
cartesian 10
Cauchy 17
— integral 17

— — theorem 103
Cauchy–Riemann equations 65
Cauchy sequence 15f

Čech cohomology 42
characteristic numbers 129f
chart 52
Chern form 149f
circular ring 49
classical Riemann–Roch 94
closed 9
— differential 104
— — form 102
coherence 77
— lemma 81
coherent sheaf 76
cohomology of sheaves 34
compact 11
— operator 17
complex 20
concrete Riemann surface 61
connected 13
countable basis 12, 14
covering degree 60
cusp 124
— sector 152

Definite 16

degree 74
— formula 154
de Rham 68
diagram chasing 22
diffeomorphism 53
differentiable map 53
differential equations 99
— form 64
dimension formula 121
direct image 135f
discontinuous 123
discrete 122
— subgroup 121f
— subset 122
— topology 122
disk 56
distinguished functions 50
divisor 73
— class 73
Dolbeault 48, 65
— complex 77
— lemma of 48, 65f, 68
— theorem of 68
dual bundle 74

Elementary symmetric 119
elliptic 122
— fixed point 127
— integrals 96
entire 147
— modular form 147
exact 20
— sequence 20
exterior derivative 64
— power 84
— product 64

Factor of automorphy 128
— presheaf 27
— sheaf 31
fibre 116
finiteness theorem 76
finite volume 121
flabby 34

formula of Stokes 70, 154
Frèchet 15
free module 72
Fuchsian group 121f
function element 61
functoriality 140
fundamental domain 121, 152
— group 57
— set 152

Gauss–Bonnet formula 151, 155
generated sheaf 31
generic point 115
geometric space 50, 52
— structure 50
global section 34, 80
Godement resolution 34
— sheaf 31, 34

Harmonic 97
— differential 97
— function 97
Hausdorff 11
hermitian form 99, 149, 152
Hirzebruch–Riemann–Roch 6
Hodge 99
— theory 99
holomorphic 52
— 1-form 67
— transformation 67
homeomorphic 10
homological algebra 20
homotopic 102
— curves 103
homotopy version 103
horocycle 125
Hurwitz 95
hyperelliptic integral 96
— Riemann surface 96

Ideal 77
induced topology 8
interior 9

Index

inverse image 136

Jacobi 97
Jacobian variety 105
Jacobi determinant 69
— inversion problem 96f, 112
— — theorem 116, 119f

Lattice 112
Laurent decomposition 49
lemma of Dolbeault 48, 65f, 68
— — Poincaré 68
Leray 45
— theorem of 45
liftable 107
line bundle 73
local automorphic form 131
locally biholomorphic 56
— compact 11
— liftable 107

Measure 152
meromorphic function 55, 118
metaplectic covering 121
— group 147
metric space 8
metrizable 16
modular form 121, 147
module 23
monstrous presheaf 31
Montel's theorem 78
Montel 17
— theorem of 17
morphism 50
multiplicity 59
multiplier system 128

Neighborhood 9
norm 16
normed space 16

One-coboundary 42
one-cocycle 42
open embedding 51
oriented 70

Parabolic 122
paracompact 13
partition of unity 14
Picard 75f
— group 75f, 79
— — of 76
piecewise smooth 102
Poincaré 65, 68
positive definite 99, 149
— semidefinite 149
presheaf 26
presheaf-exact 27
presheaf-surjective 30
principal divisor 73, 107
— ideal 77
— — ring 77
product topology 9
proper 11

Quotient space 8
— topology 8

Ramification formula 95, 143
— order 94
— point 56
refinement 13
representation 121, 145
residue 87
— map 86
— theorem 70f
restriction 26
Riemann–Hurwitz 143
— formula 95
Riemann–Roch 72
— theorem 6
— — of 72
Riemann sphere 10, 54, 124
— surface 50, 52

ring 23
— extension 24

Scalar valued 122
— — modular form 122
Schwartz 18
— theorem of 18
semi-ball 15
sequences 12
sheaf 29
sheaf-exact 31
sheafify 43
sheaf-surjective 30
short exact sequence 20
skyscraper sheaf 76
splitting principle 38
star operator 98
Stokes 69f
— formula of 69, 154
sub-presheaf 27
subsheaf 29
surface 13

Taylor expansion 132
tensor product 23
theorem of de Rham 68
— — Dolbeault 68
— — Leray 45
— — Montel 17
— — Schwartz 18
topological 10
— chart 52
— manifold 13
— map 10

— space 8
— vector space 15
topology 8
tori 60
torsion element 76
— submodule 76
total differential 104
— — form 102
— ramification order 94
transformation 66
twist 79

Unitary vector space 99
universal property 10, 23

Vector bundle 72
— valued 121
— — modular form 121
volume 152

Weak solution 109
Weierstrass 17
— theorem of 17

Zero cohomology group 34
— complex 21
— divisor 76, 83
— group 29
— order 56
— section 41, 76
— sheaf 33

www.ingramcontent.com/pod-product-compliance
Lightning Source LLC
Chambersburg PA
CBHW051524170526
45165CB00002B/599